Policy and Practices for Biodiversity in Managed Forests

Edited by Fred L. Bunnell and
Jacklyn F. Johnson

Policy and Practices for Biodiversity in Managed Forests: The Living Dance

UBC Press / Vancouver

Printed in Canada on acid-free paper ∞

ISBN 0-7748-0690-7 (hardcover)
ISBN 0-7748-0691-5 (paperback)

Canadian Cataloguing in Publication Data

Main entry under title:
Policy and practices for biodiversity in managed forests

Includes bibliographical references and index.
ISBN 0-7748-0690-7 (bound); ISBN 0-7748-0691-5 (pbk.)

1. Biological diversity conservation. 2. Forest genetic resources conservation. 3. Forest management. I. Bunnell, Fred L., 1942- II. Johnson, Jacklyn F. (Jacklyn Flett), 1939-

QH541.5.F6P64 1998	333.95'16	C98-910809-0

UBC Press gratefully acknowledges the ongoing support to its publishing program from the Canada Council for the Arts, the British Columbia Arts Council, and the Multiculturalism Program of the Department of Canadian Heritage.

Set in Stone by Artegraphica Design Co.
Printed and bound in Canada by Friesens
Copy editor: Francis J. Chow
Proofreader: Fran Aitkens
Indexer: Annette Lorek

UBC Press
University of British Columbia
2029 West Mall
Vancouver, BC V6T 1Z2
(604) 822-5959
Fax: (604) 822-6083
E-mail: info@ubcpress.ubc.ca
www.ubcpress.ubc.ca

Contents

Figures

Preface

Clark S. Binkley

Many of us who work in forestry today first encountered analytical measures of biological diversity in the seminal textbook *An Introduction to Mathematical Ecology,* authored by the renowned Canadian ecologist, E.C. Pielou.[1] In that text, she explains the various ways to measure biological diversity, ranging from counts of endemics to the Shannon-Wiener information indices.

These precise scientific measures for enumerating biological diversity apparently falter when they encounter the hot politics of contemporary forest policy and management, especially those surrounding the magnificent primary forests remaining in the coastal rain forests of the world. A simple example illustrates the point. According to the usual measures of biodiversity – genetic, species, or landscape – clearcutting half of a previously unlogged watershed would unambiguously increase the measured amount of biodiversity. But environmentalists and many other thoughtful members of the public would oppose such logging on the grounds that it would "decrease biodiversity." People who have never heard of the Shannon-Wiener index profess a great concern for biodiversity. Ecological science appears to have somehow failed the public imagination.

In the tectonics of human affairs, biological diversity lies in the subduction zone where contentious human values confront scientific uncertainty. The profession of forestry, based as it is on modernist scientism, operates poorly when these matters collide without resolution, and specifically without resolution in quantifiable measures applicable to areas the size of typical management units. Hence, this book.

Public concern for biodiversity appears to stem from deep concerns about our relationship with nature. That we struggle with this core problem of human existence is not surprising, because it constitutes one of the most ancient human concerns.

---

1 E.C. Pielou. 1969. An introduction to mathematical ecology. Wiley-Interscience, New York, NY.

Dating from 5,000 years ago, the Epic of Gilgamesh – our oldest written story – bears on this problem. King Gilgamesh of Uruk in southern Mesopotamia invented the idea of a walled city to create a place of civilized comfort and security separate from what was then the dark and terrifying wilderness.[2] But to build the wall around his city, Gilgamesh needed timber from the cedar forests that – at the time – stretched unbroken across the hills and mountains surrounding the Fertile Crescent. A ferocious monster, Humbaba, had been sent by the chief Sumerian deity Enlil to guard these forests. According to the legend, Humbaba kept the people of Uruk from harvesting the wood needed to build their stockade. The epic records that Humbaba's roar "is the storm flood," his mouth "is fire," and his breath "is death." The hero of the saga – Gilgamesh – goes into the wilderness, slays Humbaba (significantly, with an axe), and thereby makes it possible for the people of Uruk to log the forest. According to the story, once they started logging, "for two miles you could hear the sad song of the cedars." With the benefit of history, we now know the rest of the story: once the forests were gone, civilizations in the region failed and the people suffered from erratic water flows, silted-up harbours, the loss of forest-dwelling creatures, and a shortage of wood for heating, cooking, and construction.

The role of humans in nature – or perhaps should I say in the rest of nature – bedeviled the Mesopotamians, bedeviled the Greeks, and now bedevils us. The public imagination now labels this angst "biodiversity" and thereby confounds matters of science and matters of values. Separating these matters would serve our understanding of each and would provide a stronger basis for public policy and management decisions.

In a recent paper,[3] the ecologist J.P. Kimmins framed the problem well:

> Phrases such as "ecologically sound" forest harvesting or "ecologically destructive" forest management have no scientific basis and no information content outside of the context of society's prevailing value ... system. Such terms imply that the science of ecology can tell us that one particular condition of a particular ecosystem is best: that it is better than all other possible conditions which that ecosystem might be in ... These questions can only be answered in the context of ... society's prevailing preferences or value systems.

Of course, even if the human values were known with great precision and depth, conflicts over forest use would not be resolved. Uncertainty over the science – the cause-effect relationships between forest management practices

---

2 This account of the Gilgamesh epic draws heavily on John Perlin's book, *A Forest Journey* (Harvard University Press: Cambridge, MA), published in 1991.

3 J.P. Kimmins 1993. Ecology, environmentalism and green religion. Forestry Chronicle 69:285-89.

and biodiversity, however measured – is partly responsible for the current turmoil. And scientific uncertainty arises, in part, as an inevitable result of scientific epistemology.[4] To paraphrase Sir Popper, no scientific hypothesis (except a tautology) can ever be proved correct. Rather, science is the process of proving incorrect hypotheses to be false. The successive falsification of hypotheses gradually narrows the area of scientific uncertainty until finally we collectively claim to "know" something to be "true."

Too little scientific effort has been applied to many aspects of the biodiversity debate. As a consequence, many interesting and relevant hypotheses remain unfalsified. These possibly incorrect but untested hypotheses possess a measure of scientific credibility with the public. In the absence of any scientific evidence to the contrary, no credible scientific expert can refute such an hypothesis. An interest group of any political stripe can seize on a plausible but untested hypothesis that happens to fit its agenda, and claim – with complete and unassailable justification – that "there is no evidence to the contrary."

The policy debates become confused as these unfalsified but untested hypotheses compete for the limited attention of policy-makers. The policy environment is unstable because the latitude for choice is wide and the objective capacity to screen alternatives is limited. The so-called "precautionary principle" counsels inaction in the face of such scientific uncertainty, but inaction can pose major economic costs on influential sectors of society dependent on using primary forests. In the ensuing political storms, science loses its traction as a useful engine to power policy or management decisions. In the absence of science, biological realities are replaced by social constructions derived from the cultural biases or power positions of those influential in the policy process.[5]

The papers for this workshop provide some specific examples of this general problem. For example, in Chapter 5, Dan Simberloff[6] comments that "the old saw that compositionally diverse communities and ecosystems are always more stable has finally faded as ecological dogma ... at least, it is now known to be an oversimplification." Yet the popular literature on forestry abounds with such "diversity implies stability" arguments.

The efficacy of landscape corridors is a second example of having scientific uncertainty cloud our forest management practices and forest policies. In British Columbia, our new Forest Practices Code requires "forest environmental networks" as corridors of connectivity in the working landscape.

---

4 For an elaboration of this point, see C.S. Binkley. 1993. Creating a knowledge-based forest sector. Forestry Chronicle 69:294-99.
5 See W. Cronon (ed.). 1996. Uncommon ground: Rethinking the human place in nature. W.W. Norton: New York, NY.
6 D. Simberloff, "Measuring diversity of communities and ecosystems with special reference to forests," Chapter 5 of this volume.

Pathbreaking work by John Nelson at the University of British Columbia suggests that permanent landscape corridors will – just by themselves – reduce allowable annual harvest levels by 6% to 12%.[7] To put these numbers in context, these landscape corridors mean that, with all else constant, about 30,000 people will lose their jobs and the BC economy will shrink by about $2 billion per year.[8] I suspect that the public tolerance for reductions in economic activity to preserve ecological values is limited, so each hectare we devote to a corridor means a smaller area devoted to permanent preserves. Viewed in this light, are corridors the best strategy for sustaining the wild places, plants, and animals we love?

In British Columbia, the economic cost of corridors is known, is high, and is immediate. The ecological benefits may or may not be significant. There may even be ecological costs.[9] Faced with this uncertainty, many (including some of the participants here) suggest an adaptive management approach where we explicitly plan to learn as we go along. While there are credible examples in some other areas of natural resources management, do we have good examples to guide applications in forestry where the system response is slow and capital investments are more permanent?

We organized the workshop "Measuring Biological Diversity for Forest Policy and Management," which was held on 23-25 February 1994, to bring some operational resolution to these problems. This book is the result. The workshop was patterned after the famous "Dahlem Konferenzen," named after the quarter in Berlin which housed them. Organized to deal with matters of scientific uncertainty, these workshops convene a group of experts in the field to reflect on a set of papers prepared prior to the meeting. Alternating plenary sessions and working groups identify points of agreement, remaining uncertainties, and fruitful areas for further research. The

---

7 J.D. Nelson and T. Shannon. 1994. Cost and timber supply assessment of the coastal biodiversity guidelines. Contract Report prepared for the Integrated Resources Branch, BC Ministry of Forests, Victoria, BC. Over the long term, the impact of permanent corridors on Annual Allowable Cut is directly related to the amount of operable land in connectors. This ranged from a low of 6% to a high of 12% in the three case studies from coastal British Columbia. In the short term, the impact of corridors on Annual Allowable Cut depends strongly on the state of the forest at the time the corridors are imposed as management constraints. If the forest has never been logged, then the cost of retaining corridors in the landscape is low. If the forest has been extensively developed without considering landscape connectivity, then the cost may be quite high.

8 These estimates are taken from a 10% Annual Allowable Cut reduction scenario made by: G. Horne, N. Paul, and D. Riley. 1991. The provincial economic impacts of a supply reduction in the British Columbia forest sector. BC Ministry of Finance and Corporate Relations, Victoria, BC. This and other estimates of the economic effects of harvest reductions are examined in C.S. Binkley, M. Percy, W.A. Thompson, and I.B. Vertinsky. 1994. A general equilibrium analysis of the economic impact of a reduction in harvest-levels in British Columbia. Forestry Chronicle 70:449-54.

9 C.C. Mann and M.L. Plummer. 1993. The high cost of biodiversity. Science 260 (June): 1868-71.

chapters of this book comprise edited and revised versions of these plenary papers. The first chapter (written by the workshop conveners Fred Bunnell and Ann Chan-McLeod) draws together the main themes of our discussion.

We are grateful for the following organizations, listed in alphabetical order, that provided financial support for our workshop and subsequent publication of resulting material. They, of course, bear no responsibility for the analysis or conclusions presented here.

BC Ministry of Forests
Canadian Forest Service
Henry P. Kendall Foundation

Vancouver Foundation
Wildlife Habitat Canada

Clark S. Binkley
Vancouver, BC
14 June 1998

Introduction
Fred L. Bunnell

We tend to view most events occurring in and around a forest as moving slowly, even quietly. The truth is different. Forests are part of a living dance of constantly changing steps, many rapid. Dance is an appropriate metaphor. Like medicine, forestry is as much an art as a science. Among our art forms, painting is two-dimensional, sculpture three-dimensional, and dance four-dimensional, through its changes in rhythm and frequencies. The living dance in and around our forests differs from other dances primarily in the variety of rhythms involved. We can group these rhythms into three classes: the dance of life, the dance of values, and the dance of time.

The Dance of Life
Forests are alive and therefore change. They were doing so long before humans appeared on earth. Trees grew, aged, rotted, and fell down, blew over, or were burned. They still do. As they do, their three-dimensional structure changes along with their composition. The plant and animal communities within the forest respond, and species drift in and out of any specific location. In this dance of changing partners, several different rhythms are immediately evident. Over large areas, events such as the shifting of jet streams and the macroclimate help govern the frequency of windstorms, fire, and other natural disturbances. As forest structure responds to particularly emphatic beats (hurricanes, large fires), communities change as species emigrate and immigrate. But much of this migration is in response to faster rhythms, such as predation and competition. These are themselves products of still other rhythms, such as the birth and death rates that comprise demographics. And somehow, embracing all of these patterns, speciation changes the very nature of the partners in the dance. The rhythms thus range from staccato, to violent crescendos, to gentle ripples opaque to our senses but not to our histories.

This dance of life is complex. While the complexity may intrigue researchers, it is frustrating to managers. There is nothing static about a forest, so there is no tidy target that can be offered as a goal for forest management. This is especially true when maintaining biological diversity is part of the goal. Each rhythm, be it fast or slow or in between, influences biological diversity at any and every scale.

The Dance of Values

The values we attribute to forests change over time. About 2,500 years ago Druids protected sacred groves for their religious or spiritual value. Germanic laws of 1,500 years ago classified European forests according to their productivity for grazing, and recognized honey and beeswax before lumber as products of the forest. Our designation of wildlife trees is more a remembering of a value than the discovery of one anew. By 556 AD, the word *forestris* was being used in Europe to describe a tree-covered area retained to preserve hunting and fishing values.[1] Through the ages, forests have also been perceived differently by different groups. To many, forests have been impediments to movement or to settlement. Others have valued forests as habitat for fur-bearing animals or as storehouses of lumber and wood fibre. More recently, forests have come to be recognized as complex living ecosystems that provide many benefits more important than any single resource such as fur, fibre, or beeswax.

At the "Earth Summit" in Rio de Janeiro in 1992, the values of forests recognized by international agreements were expanded dramatically. Among those values, the most complex is biological diversity. It is a new value. In *Our Common Future,* the Brundtland Commission (World Commission on Environment and Development) emphasized meeting the needs of the present without impairing the future ("sustainable development"). In the description of this future, biological diversity was mentioned twice (pages 148 and 157).

The short form *biodiversity* was coined only in 1986. After the Earth Summit, the goalposts for practitioners and for policy-makers had moved. Worse, they had moved to include values that are at best ill defined and that some have termed undefinable.

We can predict little about the dance of values other than that values will continue to change, for they too are products of living things – us. What we do know is that the pace of the dance has quickened and that not all values desired of a forest are compatible with each other. Somehow policy-makers, researchers, and practitioners must choreograph the dance of values to the living dance of the forest.

---

1 References for historical uses of forests are found in: F.L. Bunnell and L.L. Kremsater. 1990. Sustaining wildlife in managed forests. Northwest Environmental Journal 6:243-69.

The Dance of Time

Forest practitioners join the dance, but their steps are constrained. The major disturbance to temperate forests is fire. Fire usually moves at two broad tempos – one, a staccato approach, alighting quickly, briefly, and leaving its mark over many small areas; another, rarer, crescendoing across thousands of hectares. Between are movements of various intensities that create a wide range of patterns. We can view fire as an erratic and possibly demented conductor, never satisfied with any particular rhythm and continually stirring up the dance. At this broad scale, forest practices follow but one step, the treatment unit. The pattern of steps across a much larger management unit may be quite intricate and include small variations defined in terms of silvicultural treatments. The forest practitioner, however, is never allowed the freedom of expression that nature simply takes as her own. The practitioner's more plodding and predictable steps attempt to fit into the other rhythms as best they can.

It is just as well that forest practitioners' steps are largely predictable, because little else is. Like a river, the dance of life is never the same. A problem, however, is that, predictable though the nature of the steps may be, their pattern across the landscape almost always takes at least two generations of practitioners to bring about. It is easy to forget past steps, their potential consequences, or even why a particularly novel movement was invoked. This would be a greater problem were the living dance not so vibrant or the goal posts stationary. Under less changeable conditions, it would be far more important to remember all the steps, just in case they worked out satisfactorily.

The practitioner's steps are usually simple to execute. Practitioners, however, face several problems. They must somehow fit each individual step or treatment into the pattern nature has trod, and they must weave the pattern of their steps through the forest is such a way that this pattern is consistent with society's values – whether sustaining timber supply or sustaining biological diversity. It attains nothing to return to a spot only to find the forest too young to meet the values. Moreover, the practitioner controls only a small portion of the rhythm. The living dance of the forest is ceaseless and unpredictable; the dance of values is similar. Because forests often grow slowly, the practitioner's dance is rarely completed. Something changes the dance along the way.

Joining the Dance

This book is about joining the dance. The dance will continue with or without our participation, but policy-makers, practitioners, and researchers are usually partners in it. We participate to enjoy, conserve, and shepherd the values we desire. To do this well, we must somehow try to match our rhythms with those of the ongoing dance.

Policy and Practices for Biodiversity in Managed Forests

1

Forestry and Biological Diversity: Elements of the Problem

Fred L. Bunnell and Ann Chan-McLeod

The living dance goes on in all ecosystems. In each system it hinders the measurement or monitoring of biological diversity (Bunnell 1994, 1998). Difficulties are especially pronounced in forested systems managed for wood fibre – the systems discussed here. This chapter provides the context for the chapters that follow. First, we review features of forested systems that define and complicate policy and management to sustain biological diversity. Then we present specific questions derived from the task of monitoring biodiversity, expanding briefly on each question. Subsequent chapters address these questions.

Forest Biodiversity: Some Defining Features

Issues surrounding biological diversity are complex in any system, but are especially so in forests. Five features of managed forests define and complicate efforts to monitor biological diversity.

(1) Forests Are Especially Rich in Biodiversity

Concern for biological diversity was stimulated first by conditions in the tropics, which arguably harbour the most diverse ecosystems in the world. Although tropical forests cover only about 7% of the earth's land surface, they contain an estimated 50-90% of earth's species (Wilson 1988; Miller and Shores 1991). Biological richness is not as apparent in temperate forests, but it is nonetheless tremendous. In Canada, forests cover about 45% of the land mass (Canadian Forest Service 1994). Boyle (1991), using data of Bunnell (1990) and Bunnell and Williams (1980), estimated that about two-thirds of the estimated 300,000 species of animals, plants, and microorganisms in Canada are forest dwelling. Although proportions vary among taxonomic groups, in any area with significant forest cover the majority of species are forest dwelling (Figure 1.1).

There is no mystery to this. Forests contain greater environmental volume than does vegetation of lesser stature. When this volume is well stratified

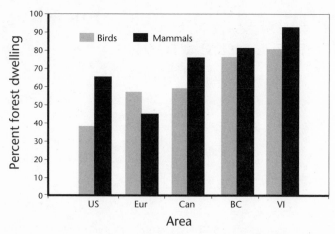

Figure 1.1   Percentage of native birds and mammals that
are forest dwelling (from Bunnell 1992). US = Continental USA;
Eur = Europe; Can = Canada; BC = British Columbia;
VI = Vancouver Island.

by trees and shrubs of different heights, more niches are available for species to use (MacArthur and MacArthur 1961; Moss 1978; Southwood et al. 1979; August 1983). The broad relationship is apparent in Figure 1.1 – the tall, well-stratified forests of western North America contain a disproportionate number of native vertebrate species. In the United States, for example, two states (Washington and Oregon) comprising less than 5% of the American land mass but containing about 25% of the conifer volume, are home to about 37% of the bird and 42% of the terrestrial mammal species native to the continental United States. Similarly, British Columbia, with about 9.5% of Canada's land mass, contains nearly 50% of the conifer volume and is home to about 72% of the native terrestrial vertebrate species (Figure 1.1).

Worldwide, natural forests are particularly rich in biological diversity, containing the majority of species on the planet. Many species are important to the continued productivity of the forest. These include such diverse organisms as mycorrhizal fungi (Trappe 1962; Trappe and Fogel 1978; Redhead 1994), the small mammals that disperse them (Fogel and Trappe 1978; Maser et al. 1978; Ure and Maser 1982), and nitrogen-fixing bacteria and their associates (Granhall 1976; MacDicken 1994). In short, forest management can affect a large portion of biological diversity, some elements of which are critical to the productivity of the forest itself.

(2) We Modify Forests to Extract Products
Forested nations rely on forests to provide products for their citizens and to generate wealth that contributes to social infrastructure and well-being. For

example, the Canadian forest industry employed 847,000 people in 1994 and provided 1 in 16 Canadian jobs (Canadian Forest Service 1995). For many forested nations, exports of forest products also contribute importantly to balance of trade and provide significant foreign revenue. In 1993, lumber, pulp, and paper sales as a percentage of total national exports equalled 34.7% for Finland, 18.0% for Sweden, 12.5% for Canada, and 12.2% for New Zealand (United Nations 1994).

Forest products are not limited to lumber and paper but are remarkably variable and include nuts, charcoal, latex, laminated beams, and alcohol. Moreover, the source of these products is renewable, and their creation is more environmentally friendly than are substitute products. Lumber, for example, is highly energy-efficient compared with other construction products, requiring only 580 kilowatts per tonne to produce. In contrast, aluminum requires about 73,080 kilowatts per tonne; plastics 3,480; and cement 2,900 (Pearson 1989).

The forests that are home to much of the world's biological diversity are simultaneously a renewable source of relatively inexpensive products that contribute importantly to human needs. Policies governing the use of forest resources must accommodate both their direct contribution to human needs and their broader contribution to the needs of other species.

(3) Modification of Forests Can Reduce Biodiversity

Conversion of forests to agriculture, suburbs, or ski slopes dramatically reduces their capacity to host all species that were present before conversion. Domestication of natural forests resulting from management can reduce biodiversity also. This reduction can happen at the stand and at the forest level.

Although any specific silvicultural treatment, including harvesting, is usually limited to stands of 1-100 ha in size, forest management distributes these stands through time and space over much larger areas of 200,000 to 1 million ha. At the stand level, forestry can locally eliminate specific structures (such as downed wood, snags, large live trees) upon which certain species depend (Aubry et al. 1988; Hunter 1990; Franklin and Spies 1991; Morrison and Raphael 1993; O'Hara et al. 1994; Dupuis et al. 1995). On a broader scale, other ecological processes enter, and populations can become isolated through lack of connectivity or can experience negative effects from too much forest edge (Fahrig and Merriam 1985; Andren and Angelstam 1988; Taylor and Fahrig 1993; Patton 1994).

Human use of forest resources will continue, as will its potential impacts on forest-dwelling biodiversity. Potential impacts in forests are magnified because forest practices influence both specific sites and the extensive areas over which these sites are distributed. Impacts may be either immediate or cumulative over decades of change in forest structure. The slow growth of forests means that the consequences of current practices may not be manifest until decades after logging or other treatment (Bunnell et al. 1993).

(4) The Public Is Genuinely Concerned about Forest-Dwelling Biodiversity

These three features – richness of forest biodiversity, modification of forests to extract products, and reduction of forest biodiversity – have combined to create genuine public concern. The World Commission on Environment and Development (Brundtland Commission) of 1987 acknowledged these concerns and offered the concept of sustainable development to express a balance between economic and ecological tensions. The commission mentioned biological diversity (pages 148 and 157 of *Our Common Future*) but did not address the complexities and obstacles facing its conservation. Concern did not diminish with the Brundtland Commission, and within five years of its report, several major international agreements were signed at the United Nations Conference on Environment and Development (UNCED 1992), or "Earth Summit," in Rio de Janeiro. Four of these agreements directly address how forests are managed, and three explicitly include biological diversity.

The Convention on Biological Diversity contains three major national obligations: (1) to conserve biological diversity, (2) to use biological resources in a sustainable manner, and (3) to share the benefits of biodiversity fairly and equitably. *Agenda 21* addresses the conservation and rational use of forests and the conservation of biodiversity. *Guiding Principles on Forests*[1] specifically addresses forest conservation and the integration of forest management with adjacent areas to maintain ecological balance and sustainable productivity. The *Framework Convention on Climate Change* includes the goals of promoting sustainable forest management and conserving and enhancing carbon sinks (such as forests and wetlands).

This rapid expansion of the scope of international agreements addressing forestry could not have occurred without widespread and deeply felt public concern. With such rapid expansion came associated problems. First, public concerns had rapidly outstripped managers' experience in meeting those concerns. Second, the perceived urgency led to terms within agreements that captured the range of concerns but were ill defined, such as *biological diversity* or *holistic forestry*. As stated in the discussion of "ecosystem management" in the 1993 Congressional Research Report (United States General Accounting Office 1994), "There is not enough agreement on the meaning of this concept to hinder its popularity." Policy had outpaced practical experience and theoretical frameworks for practice (see Chapter 4 of this volume).

---

1   The full title is *Non-legally Binding Authoritative Statement of Principles for a Global Consensus on the Management, Conservation, and Sustainable Development of All Types of Forest.*

(5) Many Principles Underlying Conservation of Biodiversity Are Novel

While it is clear that concern and policy can outpace practice, it is less apparent how theory can be outpaced. Two features of biological diversity delayed the advancement of theory: the number of entities involved and the scales of processes involved. Definitions of biological diversity include all living things and the variability among them (Bunnell 1994; Delong 1996). We do not know the number of species encompassed by biological diversity, but estimates range from about 5 to 30 million species (Wilson 1988). Many of these have been neither sampled nor named. Processes acting to sustain or eliminate species occur over a large range of spatial and temporal scales (see Chapter 7). Not only are much larger scales of influence newly recognized as important (such as through fragmentation), but the simultaneous interaction of processes across several spatial and temporal scales is considered important (Wiens et al. 1986; Bunnell 1992; Noss 1992; Bunnell and Huggard 1998). Potential consequences in forests are especially obvious because extracting forest products simultaneously influences habitat at very different levels: from individual harvest or treatment units to the distribution of these units across hundreds of thousands of hectares over decades of practice.

Seminal references for theory relevant to broad-scale processes are recent, such as MacArthur and Wilson (1967) for island biogeography and Levins (1970) for metapopulations. Although researchers and theoreticians have debated mightily over concepts derived from these sources, and over the utility of the concepts in guiding practice (e.g., Simberloff and Abele 1976a, 1976b, 1982; Gilbert 1980; Simberloff 1986), in truth there has been little time for practitioners to apply and evaluate the concepts (Bunnell and Dupuis 1995). Policy changes, however, have led to their application in forested systems as the debate continues. As a result, research, training, and practice are uncommonly disconnected (e.g., Binkley 1992; Namkoong 1993). Again, the potential problem is magnified in forestry because everyday management decisions have long-lasting consequences.

In summary, forests are particularly rich in biodiversity, and our continued use of forests can dramatically affect forest-dwelling biodiversity. The public is thus justifiably concerned about issues it perceives to be embodied in the concept of biological diversity, but *policy created to sustain biological diversity is well ahead of research and understanding.*

Issues in Measuring Biodiversity for Forest Policy and Management

Governments have committed to sustain biodiversity (Bunnell 1994; Galindo-Leal and Bunnell 1995) but have found it difficult to create operational goals and to demonstrate progress (Delong 1996). Tactics to maintain portions of biodiversity in managed forests are not lacking (reviews of Freedman et al. 1994; Bunnell et al. 1997; Bunnell and Kremsater 1998). The major obstacle to formulating strategic goals and evaluating progress has been a lack of agreement about defining, measuring, and monitoring the sum of biodiversity (Rochelle and Hicks 1996; Bunnell 1998). This book summarizes our understanding about defining, measuring, and monitoring biodiversity as these relate to forest policy and management.

For any management task, practitioners must address three broad challenges:

- Where are we going? (What is the target?)
- How do we get there? (What are our tactics?)
- How will we know when we are there? (How do we monitor success?)

The first challenge is strategic and recognizes that the activities of management and policy have some goal. The second is tactical and defines the kinds of activities that will attain that goal. The third is evaluative and acknowledges that some measure of success is necessary. The first and third challenges are necessarily related: measures of success must be related to the goal. Many tactical questions are site-specific (such as whether to create snags or retain wildlife trees), but some relate directly to goal setting or monitoring (such as size of planning units). Below we summarize 17 questions related to measuring and monitoring biological diversity, grouped under the three broad challenges faced by managers and policy-makers. These questions arose from the workshop (see Preface) that formed the core of this book.

Where Are We Going?
Bunnell (1994, 1998) and Delong (1996) reviewed almost 90 definitions of biological diversity from documents written to guide policy and management decisions. Almost all definitions included a phrase similar to the following: "the full range of life in all its natural forms, including genes, species and ecosystems – and the ecological processes that link them." In terms of goal setting, this apparent agreement on the content of biological diversity is illusory: it belies the enormous difficulties of translating definitions into an operational statement that can be used by land managers and policy-makers to guide action and legislation.

Major questions associated with goal setting include the following.

(1) What Is the Goal?
The scope of biological diversity yields a remarkably complex, largely unknown, and ill-defined goal. Among the entities embraced by biological diversity, only species are easy to envision. Most species, however, are unknown and have not yet been described. Other entities (genes, processes,

communities, or ecosystems) are either "invisible" or extremely difficult to define, or are concepts rather than entities. Targeting all biological entities embraced by existing definitions inevitably leads to a remarkably complex and largely unknowable goal. The relative importance of various components of biological diversity is undefined. It is likewise unclear how to integrate and manage the components that should be included.

(2) What Role Do Values and Perspectives Play in Defining the Goal?
It is important for policy-makers to determine which concerns about biological diversity are primarily value-driven rather than securely based on research findings. There are two broad issues. One is the outcomes desired and valued from specific management actions. The second is the relative emphasis placed on components embraced by the concept of biodiversity. The relative emphasis placed on such components can be expressed as further questions.

Are all features equal? Definitions of biological diversity embrace a wide range of features of the natural world, but rationales for conserving biological diversity do not consistently emphasize the same features (e.g., Delong 1996; Bunnell 1998). A rationale may be derived largely from practical considerations or from moral and aesthetic values (Ehrenfeld 1988; Bunnell 1990). Rationales based primarily on practical values tend to emphasize natural features such as manipulable structures, recognizable habitats, and processes believed to sustain the regenerative capacities of ecosystems. Rationales emphasizing aesthetic values tend to emphasize attractive structural features such as large trees ("cathedral" values) or species deemed attractive or simply rare.

Are all species equal? Writers on biological diversity have variously stressed the importance of conserving all species, implying equality among species (e.g., Wilson 1988; McNeely et al. 1990), or suggested a focus on particular species, such as species occurring in small numbers or at few locations (e.g., Arita et al. 1990). Others have argued for placing greater conservation priorities on species that play known, critical ecological roles rather than on species with apparently larger degrees of functional redundancy (Walker 1992, 1995). Analogous questions apply to the equality of entities at other levels of integration: do rare alleles, communities, or ecosystems merit special attention in the same way as has been suggested for rare species?

Are all places equal? Definitions of biological diversity provide no guidance on whether policy should emphasize conservation on a local, regional, national, or global basis. Populations at the periphery of a species' range, just entering one political unit but abundant in adjacent units, provide an obvious quandary. The importance of sustaining peripheral populations is unclear. Is it important, for example, to conserve existing distributions?

What risks should we take?   All management involves risk (Bunnell 1976; Walters 1986). Ideally, specific actions taken to conserve biological diversity should be subject to an assessment of the benefits and costs of taking action relative to the risks of doing nothing. Can we, for example, specify population sizes at which changes in management actions are necessary? In viable-population analysis (Soulé 1987), what should be the acceptable time frame for probable extinction, and at what probability? Are these parameters universal, or should they differ among species, genes, or ecosystems?

(3) Over What Area and Time Period Should the Goal Be Defined?
A definition that crosses scales in level of biological integration (i.e., from genes to ecosystems) must also cross scales in both time and space. The appropriate temporal and spatial scales for managing shrews, for example, are woefully inadequate for managing grizzly bears. Likewise, components of biological diversity naturally inhabit different portions of any large area as time passes and natural succession occurs (Bunnell 1992). There are few clear natural boundaries to the size of area or period of time for which a target for biological diversity can be defined.

(4) What Is Ecosystem Integrity?
Most definitions of biological diversity include the notion of ecosystem integrity, often by mentioning processes. Currently, we have no agreed-upon definition of ecological or ecosystem integrity, especially for terrestrial systems (National Academy of Sciences 1986). In fact, in most definitions of biological diversity, the methods for measuring ecosystem function are not addressed; the work of the Scientific Panel for Sustainable Forest Practices in Clayoquot Sound (Clayoquot Scientific Panel) is an exception (Scientific Panel 1995).

(5) Can Indices of Ecological Diversity Help Define Goals?
Many authors argue that simple richness (number of species in an area) provides an inadequate assessment of biological diversity (e.g., Simpson 1949; Magurran 1988; Usher and Pineda 1991; Colwell and Coddington 1994). Numerous indices have been devised that are based on measures (or estimates) of the quantities of different species in each sample plot (the two best known are Shannon's Index and various inverse functions of Simpson's Index of Concentration). Other statistical descriptors of ecological diversity (such as alpha, beta, and gamma diversity) have also been invoked to suggest management tactics (e.g., Namkoong 1991; Pimm and Gittleman 1992; Samson and Knopf 1993).

Diversity indices may not be helpful for two broad reasons: (1) their statistical formulation commonly emphasizes evenness (Patil and Taille 1982), and (2) they treat all species as equal (Pielou 1995). While high evenness may lead us to perceive the community to be diverse, this condition is an uncommon and unnatural phenomenon; similarly, many workers do not ascribe equal weight to ubiquitous and rare species (Bunnell 1998). Alpha and beta diversity are similarly suspect because: (1) they are incommensurate (alpha diversity is the number of species in an area, such as a stand; beta diversity is the rate of change in species between one area or stand and the next),[2] and (2) increases in beta diversity may reflect only the responses of mobile, ubiquitous species (Pielou 1995).

(6) How Different Does an Entity Have to Be Before It Contributes to Biodiversity?
The Convention on Biological Diversity specifically emphasizes variability among living organisms and distinguishes between this variability (defined as biological diversity) and biological resources or entities (such as species). The amount of variation sufficient to generate diversity is poorly understood. Except for cases of parthenogenesis (development of an egg without fertilization) and identical twinning, virtually no two members of a vertebrate species are genetically identical (Selander 1976); thus each individual is a candidate for conservation. More arbitrarily, we can "create" greater diversity simply by subdividing categories. Thus, forest structure as represented by an age class of 40-80 years can be divided into two age classes aged 40-60 years and 60-80 years; a portion of structural diversity ostensibly doubles.

(7) Is Viability (Large Tracts) or Representation (Several Smaller Tracts) the Dominant Goal of Protected Areas?
Protected areas or tracts are part of any conservation strategy. The goal of minimizing extinction rates or maintaining viability is better served by protecting large tracts (Diamond 1975); the goal of representing each recognizably distinct example (see question 6) of different communities may be better served by several smaller tracts (Connor and Simberloff 1984; Samson and Knopf 1993). The appropriate emphases on viability and on representation are unclear.

How Do We Get There?
Any operational goal for sustaining biological diversity will be complex and uncertain. Both the strategy of policy and the tactics of management must encompass that uncertainty and complexity.

---

2 There are at least six ways of quantifying beta diversity (e.g., Wilson and Shmida 1984; Magurran 1988), but the major distinction from alpha diversity remains the same.

(8) How Do We Develop Strategy or Tactics When the Target Is Poorly Defined and the Scientific Information Is Largely Incomplete?
Despite the uncertainties summarized in the preceding seven questions, practitioners must proceed with actions now. These actions should not preclude future, more desirable options. It is especially unclear how we can integrate management of biodiversity at the genetic, population, community, and landscape levels. Likewise, how can we integrate the compositional, structural, and functional aspects of biodiversity that are emphasized in various definitions?

(9) Given That the Target Moves, How Can We Develop Appropriate Tactics?
Distributions of genetic diversity across the landscape change with time, as do the distributions of species. Genetic variation changes as a function of demography and much longer term processes such as isolation and genetic drift; distributions of species change naturally with succession and more gradually with climate change (e.g., Hebda and Whitlock 1997). Tactics for maintaining biological diversity must accommodate a moving target (Bunnell et al. 1993).

(10) What Are Appropriate Planning Units and Time Horizons?
Planning scales for biodiversity are ill defined because: (1) the empirical species-area curve is continuous; (2) legal boundaries do not coincide with ecological units; and (3) in metapopulations, the flow of individuals and genetic materials occurs among otherwise recognizably discrete biological units.

Large areas are necessary to accommodate all seral stages and their associated species as well as wide-ranging species (Bunnell and Kremsater 1990; Bunnell 1992). Provision of habitat, however, must recognize finer scales because requirements vary dramatically among species and seasonally within species. No single spatial scale is sufficient. Managing for biodiversity also must recognize several time scales because of the variety of processes that can influence the outcome of any practice. Population dynamics, natural succession, and long-term genetic processes all have potential effects, and each occurs at very different rates (Bunnell and Huggard 1998). Forest practices themselves have effects at different scales and rates.

(11) What Are the Relative Roles of Protected Areas and Managed Forests in Biological Conservation?
Creating forested protected areas decreases the area from which timber is harvested and usually decreases the revenue used to maintain social infrastructure. Policies to maintain biological diversity within managed forests also reduce the amounts of fibre potentially harvested. It is thus important to assess the relative roles these two broad approaches play in sustaining biological diversity.

(12) Should Managed-Forest Mosaics Mimic Those Generated by Natural Disturbance Regimes?
A growing body of literature and public statements argues that biological diversity would be better sustained if forest practices created forests that more closely resembled those produced under natural disturbance regimes (Pickett and Thompson 1978; White and Bratton 1980; Harris 1984; Bunnell and Kremsater 1990; Hunter 1990). Is there evidence that management tactics should mimic natural disturbance regimes?

(13) Can We Assess the Relative Impacts of Efforts Devoted to Stand-Level and Forest-Level Activities?
Forest practices at both the stand (Spies and Cline 1988; Morrison and Raphael 1993; Dupuis et al. 1995) and forest level (Gates and Gysel 1978; Askins et al. 1987) can modify biological diversity. Different tactical and strategic questions arise at each level. Very broadly, the major questions can be phrased as follows: How much of what (stand or habitat) is enough? How should that be distributed (forest or landscape)? Research should focus on reducing uncertainty, and actions should be taken where they have the most effect.

(14) Can Diversity Indices Help Define Tactics?
Strategies to maximize alpha, beta, and gamma diversity may conflict with each other. For example, alpha diversity may be maximized by maintaining a large uneven-aged tract of forest. In contrast, beta diversity may be maximized by dividing the same forest into young, intermediate, and old even-aged stands. Which of these concepts, if any, are relevant to management tactics?

How Will We Know When We Get There?
Given the apparent urgency of maintaining biological diversity (Soulé 1985; Wilson 1988; McNeely et al. 1990) and the legal obligations flowing from the Convention on Biological Diversity, some evaluative process is essential. Three broad questions arise.

(15) What Do We Monitor?
Given the scope of elements and processes included within definitions of biological diversity – genes, species, ecosystems, structures, processes – what should be monitored to evaluate relative success? Species are the most tractable element, but they do not represent the variability emphasized within the Convention on Biological Diversity. It is also logistically impossible to monitor all species individually. The use of keystone, umbrella, or indicator species presents its own problems (Ryti 1992). Can a small number of species adequately represent a diverse fauna with dramatically

different natural histories? Are communities or species assemblages permanent units, and what do we do if they are not? Are there surrogate variables, such as snags or downed wood, that are reliable indicators?

(16) How Do We Monitor?
If we monitor components of current definitions of biological diversity, there are difficulties at each level of biological organization (many have already been noted). The difficulties are clearly exposed at the fundamental genetic level. Most measurement techniques are complex, costly, and labour-intensive, and may not provide meaningful measures of potentially important biological responses such as inbreeding depression. Furthermore, we have no known target or goal. Even if completely censused, existing conditions do not represent some "ideal" genetic structure but a "snapshot" of conditions known to be dynamic (Bunnell et al. 1991). We are hindered by logistical problems in identifying metapopulations and dispersal or migration patterns. For most populations, only a very small proportion of individuals disperse long distances, but it is the infrequent but far-ranging dispersal events that have large implications for gene flow (Bunnell and Harestad 1983; Cockburn 1992). Simple monitoring approaches applicable to large areas are undefined.

(17) How Do We Make Adaptive Management Work?
Should policy decisions be implemented in the form of rigorously designed management experiments that enforce both monitoring and research? How do we extend authority and flexibility to modify existing guidelines if these prove inappropriate?

These questions were raised before and during the workshop. This book does not unequivocally answer each of them, but addresses many in the chapters that follow. Chapter 2 provides a broad overview of issues surrounding biological diversity in forests, primarily from the perspective of policy-makers. Subsequent chapters treat different levels of biological organization, beginning with genetics (Chapter 3), through populations (Chapter 4) and communities or ecosystems (Chapter 5), to broad landscape units such as landforms (Chapter 6). Chapter 7 then addresses the issue of appropriate scale for management of biological diversity. Because concepts related to biological diversity cross scales and levels of integration, different chapters offer different perspectives on the 17 specific questions asked above. These perspectives are summarized in the final chapter (Chapter 8), and tentative answers are provided for each question.

Literature Cited
Andren, H., and P. Angelstam. 1988. Elevated predation rates as an edge effect in habitat islands: Experimental evidence. Ecology 69:544-47.

Arita, H.T., J.G. Robinson, and K.H. Redford. 1990. Rarity in neotropical forest mammals and its ecological correlates. Conservation Biology 4:181-82.

Askins, R.A., M.J. Philbrick, and D.S. Sugeno. 1987. Relationship between the regional abundance of forest and the composition of forest bird communities. Biological Conservation 39:129-52.

Aubry, K.B., L.L.C. Jones, and P.A. Hall. 1988. Use of woody debris by plethodontid salamanders in Douglas-fir forests in Washington. Pp. 32-37 *in* R. Szaro, K.E. Severson, and D.R. Patton (tech. coords.). Proceedings of a symposium on management of amphibians, reptiles, and small mammals in North America, 19-21 July 1988, Flagstaff, AZ. General Technical Report RM-166, US Department of Agriculture (USDA) Forest Service, Rocky Mountain Forest and Range Experiment Station, Fort Collins, CO.

August, P.V. 1983. The role of habitat complexity and heterogeneity in structuring tropical mammal communities. Ecology 64:1495-507.

Binkley, C.S. 1992. Forestry after the end of nature. Journal of Forestry 90:33-37.

Boyle, T.S.B. 1991. Biodiversity of Canadian forests: Current status and future challenges. Forestry Chronicle 68:444-52.

Bunnell, F.L. 1976. The myth of the omniscient forester. Forestry Chronicle 52:150-52.

–. 1990. Biodiversity: What, where, why, and how. Pp. 29-45 *in* The silviculture conference: Stewardship in the new forest, 18-20 November 1991, Vancouver, BC. Forestry Canada, Ottawa, ON.

–. 1992. De mo' beta blues: Coping with the landscape. Pp. 45-58 *in* Proceedings of a seminar on integrated resource management, Forestry Canada, Maritimes Region. Information Report M-X-183E/F.

–. 1994. Toto, this isn't Kansas: Changes in integrated forest management. Pp. 1-14 *in* Selected papers from the Integrated Forest Management Workshop, Proceedings of the Canadian Model Forest Network Workshop, 3-5 October 1994, Sussex, New Brunswick. Canadian Forest Service, Ottawa, ON.

–. 1998. Overcoming paralysis by complexity when establishing operational goals for biodiversity. Journal of Sustainable Forestry 7(3/4):145-64.

Bunnell, F.L., and L.A. Dupuis. 1995. *Conservation Biology's* literature revisited: Wine or vinaigrette? Wildlife Society Bulletin 23(1):56-62.

Bunnell, F.L., and A.S. Harestad. 1983. Dispersal and dispersion of black-tailed deer – models and observations. Journal of Mammalogy 64:201-9.

Bunnell, F.L., and D.J. Huggard. 1998. Biodiversity across spatial and temporal scales: Problems and opportunities. Forest Ecology and Management. In press.

Bunnell, F.L., and L.L. Kremsater. 1990. Sustaining wildlife in managed forests. Northwest Environmental Journal 6:243-69.

–. 1994. Tactics for maintaining biodiversity in forested ecosystems. Proceedings of the XXI International Union of Game Biologists (IUGB) Congress 1:62-72.

–. 1998. Managing to sustain vertebrate diversity in forests of the Pacific Northwest: Relationships within stands. Environmental Review. In press.

Bunnell, F.L., and R.G. Williams. 1980. Subspecies and diversity – the spice of life or prophet of doom. Pp. 246-59 *in* R. Stace-Smith, L. Johns, and P. Joslin (eds.). Threatened and endangered species and habitats in British Columbia and the Yukon. BC Ministry of the Environment, Fish and Wildlife Branch, Victoria, BC.

Bunnell, F.L., D.K. Daust, W. Klenner, L.L. Kremsater, and R.K. McCann. 1991. Managing for biodiversity in forested ecosystems. Report to Forest Sector of the Old-Growth Strategy, Centre for Applied Conservation Biology, University of British Columbia, Vancouver, BC.

Bunnell, F.L., C.G. Galindo-Leal, and J. Nelson. 1993. Ecological restoration in forested landscapes: Problems and opportunities (British Columbia). Restoration and Management Notes 11:56-57.

Bunnell, F.L., L.L. Kremsater, and R.W. Wells. 1997. Likely consequences of forest management on terrestrial, forest-dwelling vertebrates in Oregon. Oregon Forest Resources Institute, Portland, OR.

Canadian Forest Service. 1994. Forest profiles: Focus on the players. Chapter 5 *in* Canada's forests 1993: Forests, a global resource. Fourth Report to Parliament, Minister of Supply and Services Canada.

–. 1995. The state of Canada's forests: A balancing act. Canadian Forest Service, Natural Resources Canada, Ottawa, ON.

Cockburn, A. 1992. Habitat heterogeneity and dispersal: Environmental and genetic patchiness. Chapter 4 *in* Nils Chr. Stenseth and W.Z. Lidicker, Jr. (eds.). Animal dispersal – small mammals as a model. Chapman and Hall, New York, NY.

Colwell, R.K., and J.A. Coddington. 1994. Estimating the extent of terrestrial biodiversity through extrapolation. Philosophical Transactions of the Royal Society of London B 345:101-18.

Connor, E.F., and D. Simberloff. 1984. Neutral models of species co-occurrence patterns. Pp. 316-31 *in* D.R. Strong, D. Simberloff, L.G. Abele, and A.B. Thistle. Ecological communities. Princeton University Press, Princeton, NJ.

Delong, D.C. 1996. Defining biodiversity. Wildlife Society Bulletin 24:738-49.

Diamond, J.M. 1975. The island dilemma: Lessons of modern biogeographical studies for the design of natural reserves. Biological Conservation 7:129-46.

Dupuis, L.A., J.N.M. Smith, and F.L. Bunnell. 1995. Relation of terrestrial amphibian abundance to tree-stand age. Conservation Biology 9:645-53.

Ehrenfeld, D. 1988. Why put a value on biodiversity? Pp. 212-16 *in* E.O. Wilson and F.M. Peter (eds.). Biodiversity. National Academy Press, Washington, DC.

Fahrig, L., and G. Merriam. 1985. Habitat patch connectivity and population survival. Ecology 66:1762-68.

Fogel, R.M., and J.M. Trappe. 1978. Fungus consumption (mycophagy) by small animals. Northwest Science 52:1-31.

Franklin, J.F., and T.A. Spies. 1991. Composition, function, and structure of old-growth Douglas-fir forests. Pp. 71-80 *in* L.F. Ruggiero, K.B. Aubry, A.B. Carey, and M.H. Huff (tech. coords.). Wildlife and vegetation of unmanaged Douglas-fir forests. General Technical Report PNW-GTR-285, US Department of Agriculture (USDA) Forest Service, Pacific Northwest Research Station, Portland, OR.

Freedman, B., S. Woodley, and J. Loo. 1994. Forest practices and biodiversity with particular reference to the Maritime Provinces of eastern Canada. Environmental Review 2:33-77.

Galindo-Leal, C., and F.L. Bunnell. 1995. Ecosystem management: Implications and opportunities of a new paradigm. Forestry Chronicle 71:601-6.

Gates, J.E., and L.W. Gysel. 1978. Avian nest dispersion and fledging success in field forest ecotones. Ecology 59:871-83.

Gilbert, F.S. 1980. The equilibrium theory of biogeography: Fact or fiction? Journal of Biogeography 7:209-35.

Granhall, U. (ed.). 1976. Environmental role of nitrogen-fixing blue-green algae and asymbiotic bacteria. Ecological Bulletins (Stockholm), Vol. 26.

Harris, L.D. 1984. The fragmented forest: Island biogeography theory and the preservation of biotic diversity. University of Chicago Press, Chicago, IL.

Hebda, R.J., and C. Whitlock. 1997. Environmental history. Pp. 227-54 *in* P.K. Schoonmaker, B. von Hagen, and E.C. Wolf (eds.). The rain forests of home. Island Press, Washington, DC.

Hunter, M.L., Jr. 1990. Wildlife, forests, and forestry. Prentice Hall, Englewood Cliffs, NJ.

Levins, R. 1970. Extinction. Pp. 77-107 *in* M. Gerstenhaber (ed.). Lectures on mathematics in the life sciences, Vol. 2. American Mathematical Society, Providence, RI.

MacArthur, R.H., and J.W. MacArthur. 1961. On bird species diversity. Ecology 42:594-98.

MacArthur, R.H., and E.O. Wilson. 1967. The theory of island biogeography. Princeton University Press, Princeton, NJ.

MacDicken, K.G. 1994. Selection and management of nitrogen-fixing trees. Winrock International Institute for Agricultural Development, Morrilton, AR.

Magurran, A.E. 1988. Ecological diversity and its measurement. Princeton University Press, Princeton, NJ.

Maser, C., J.M. Trappe, and R.A. Nussbaum. 1978. Fungal-small mammal interrelationships with emphasis on Oregon coniferous forests. Ecology 59:799-809.

McNeely, J.A., K.R. Miller, W.V. Reid, R.A. Mittlemeier, and T.B. Werner. 1990. Strategies for conserving biodiversity. Environment 32:16-40.

Miller, K.R., and J.N. Shores. 1991. Biodiversity and the forestry profession. H.R. MacMillan Lecture Series, No. 41. University of British Columbia, Vancouver, BC.

Morrison, M.L., and M.G. Raphael. 1993. Modeling the dynamics of snags. Ecological Applications 3:322-30.

Moss, D. 1978. Diversity of woodland song-bird populations. Journal of Animal Ecology 47:521-27.

Namkoong, G. 1991. Biodiversity issues in genetics, forestry, and ethics. Forestry Chronicle 68:438-43.

–. 1993. Integrating science and management at the University of British Columbia. Journal of Forestry 91:24-27.

National Academy of Sciences. 1986. Ecological knowledge and environmental problem-solving: Concepts and case studies. National Academy Press, Washington, DC.

Noss, R.F. 1992. Issues of scale in conservation biology. *In* P.L. Fieldler and S.K. Jain (eds.). Conservation biology: The theory and practice of nature conservation, preservation, and management. Chapman and Hall, New York, NY.

O'Hara, K.L., R.S. Seymour, S.D. Tesch, and J.M. Guldin. 1994. Silviculture and our changing profession. Leadership for shifting paradigms. Journal of Forestry 92:8-13.

Patil, G.P., and C. Taille. 1982. Diversity as a concept and its measurement. Journal of the American Statistical Association 77(379):548-61.

Patton, P.W.C. 1994. The effect of edge on avian nest success: How strong is the evidence? Conservation Biology 8:17-26.

Pearson, D. 1989. The natural house book: Creating a healthy, harmonious, and ecologically sound home environment. Simon and Schuster, New York, NY.

Pickett, S.T.A., and J.N. Thompson. 1978. Patch dynamics and the design of nature reserves. Biological Conservation 13:27-37.

Pielou, E.C. 1995. Biodiversity versus old-style diversity: Measuring biodiversity for conservation. Chapter 2 *in* T.J.B. Boyle and B. Boontawee (eds.). Measuring and monitoring biodiversity in tropical and temperate forests. Center for International Forestry Research (CIFOR), Bogor, Indonesia.

Pimm, S.L., and J.L. Gittleman. 1992. Biological diversity: Where is it? Science 255: 940.

Redhead, S.A. 1994. Macrofungi of British Columbia. Pp. 81-89 *in* L.E. Harding and E. McCullum (eds.). Biodiversity in British Columbia: Our changing environment. Environment Canada, Canadian Wildlife Service, Pacific and Yukon Region.

Rochelle, J.A., and L.L. Hicks. 1996. The role of private industrial forest lands in the management of biological diversity. Pp. 655-63 *in* R.C. Szaro and D.W. Johnston (eds.). Biodiversity in managed landscapes. Oxford University Press, New York, NY.

Ryti, R.T. 1992. Effect of the focal taxon on the selection of nature reserves. Ecological Applications 2:404-10.

Samson, F.B., and F.L. Knopf. 1993. Managing biodiversity. Wildlife Society Bulletin 21:509-14.

Scientific Panel for Sustainable Forest Practices in Clayoquot Sound. 1995. Report 5, Sustainable ecosystem management in Clayoquot Sound: Planning and practices. Ministry of Environment, Lands and Parks, Victoria, BC.

Selander, R.K. 1976. Genetic variation in natural populations. Pp. 21-45 *in* F.J. Ayala (ed.). Molecular evolution. Sinauer Associates, Sunderland, MA.

Simberloff, D.S. 1986. Design of nature reserves. Pp. 315-17 *in* M.B. Usher (ed.). Wildlife conservation evaluation. Chapman Hall, London, UK.

Simberloff, D.S., and L.G. Abele. 1976a. Island biogeography theory and conservation practice. Science 191:285-86.

–. 1976b. Island biogeography and conservation strategy and limitations. Science 193:1027-29.

–. 1982. Refuge design and island biogeographic theory: Effects of fragmentation. American Naturalist 120:41-50.

Simpson, E.H. 1949. Measurement of diversity. Nature 163:88.

Soulé, M.E. 1985. What is conservation biology? BioScience 35:727-34.

– (ed.). 1987. Viable populations for conservation. Cambridge University Press, New York, NY.

Southwood, T.R.E., V.K. Brown, and P.M. Reader. 1979. The relationships of plant and insect diversities in succession. Biological Journal of the Linnean Society 12:327-48.

Spies, T.A., and S.P. Cline. 1988. Coarse woody debris in forests and plantations of coastal Oregon. Pp. 5-23 *in* C. Maser, R.F. Tarrant, J.M. Trappe, and J.F. Franklin (eds.). From forest to the sea: A story of fallen trees. General Technical Report PNW-GTR-229, US Department of Agriculture (USDA) Forest Service, Portland, OR.

Taylor, P.D., and L. Fahrig. 1993. Connectivity is a vital element of landscape structure. Oikos 68:571-73.

Trappe, J.M. 1962. Fungus associates of ectotrophic mycorrhizae. Botanical Review 28:538-606.

Trappe, J.M., and R. Fogel. 1978. Ecosystematic functions of mycorrhizae. Pp. 205-14 *in* J. Marshall (ed.). The belowground symposium: A synthesis of plant-associated processes. Range Sciences Series, No. 27. Colorado State University, Fort Collins, CO.

United Nations. 1994. Food and Agricultural Organization (FAO) trade yearbook, 1993. Food and Agricultural Organization (FAO) Statistics Series, No. 121. Food and Agricultural Organization of the United Nations, Rome.

United States General Accounting Office. 1994. Ecosystem management: Additional actions needed to adequately test a promising approach. GAO/RECD-94-111, US General Accounting Office, Washington, DC.

Ure, D.C., and C. Maser. 1982. Mycophagy of red-backed voles in Oregon and Washington. Canadian Journal of Zoology 60:3307-15.

Usher, M.B., and F.D. Pineda (eds.). 1991. Biological diversity. Fundacion Ramon Areces, Madrid.

Walker, B.H. 1992. Biodiversity and ecological redundancy. Conservation Biology 6(1):18-23.

–. 1995. Conserving biological diversity through ecosystem resilience. Conservation Biology 9(4):747-52.

Walters, C.J. 1986. Adaptive management of renewable resources. Macmillan Publishing, New York, NY.

White, P.S., and S.P. Bratton. 1980. After preservation: Philosophical and practical problems of change. Biological Conservation 18:241-55.

Wiens, J.A., J.F. Addicott, T.J. Case, and J. Diamond. 1986. Overview: The importance of spatial and temporal scale in ecological investigations. Pp. 145-53 *in* J. Diamond and T.J. Case (eds.). Community ecology. Harper and Row, New York, NY.

Wilson, E.O. 1988. The current state of biological diversity. Pp. 3-18 *in* E.O. Wilson and F.M. Peter (eds.). Biodiversity. National Academy Press, Washington, DC.

Wilson, M.V., and A. Shmida. 1984. Measuring beta diversity with presence-absence data. Journal of Ecology 72:1055-64.

World Commission on Environment and Development (Brundtland Commission). 1987. Our common future. Oxford University Press, New York, NY.

2
Forest Policy, Management, and Biodiversity
Jagmohan S. Maini

Introduction

Conservation and sustainable development of forests have emerged as high priorities on international political, policy, science, environment, trade, and socio-economic agendas. Efforts to conserve, monitor, and judiciously use forest biodiversity are recognized as basic elements of practising sustainable forestry. Canadian forest policies and forest management practices that embrace biodiversity must be scientifically based but must also recognize the complex domestic, trans-boundary, and global dimensions of the issue. Our knowledge of the structure and function of forest biodiversity in Canada is inadequate, and we have limited resources for research and monitoring in relation to the amount of forest land in Canada. A range of approaches to conserving biodiversity in situ and ex situ can be considered. The policy on forest biodiversity should be clear, robust, and based on our best available knowledge; it should also embrace the precautionary principle. Research and monitoring must be an integral component of the policy, which should also include appropriate feedback so that we can make the necessary adjustments to policy and to regulations governing forest management.

Context

Global Perspectives

During the past 30 years, the agendas of the environmental, scientific, policy, and political communities have been evolving from local to national, regional, and global scales. In the late 1960s and early 1970s, people were concerned about local pollution of water and air. In the mid-1970s and early 1980s, they were preoccupied with meso-scale and trans-boundary issues such as acid rain and the health of the Great Lakes. During the past 10 years, the scale, scope, and complexity of people's concerns has shifted to global issues such as climate change, ozone depletion, the health of oceans and the biosphere, land degradation, loss of biodiversity, "global change," and "sustainable development." This trend towards the globalization of environmental issues is associated with globalization in a number of other areas, such as manufacturing, marketing, investments, and information.

During the past 30 years, forest-related issues have followed a similar path towards globalization. The following developments are particularly noteworthy:

- The emergence of green consumerism and the interface between international trade and the environment is of particular interest to the forestry community. The "green signals" from the marketplace demand "green" forest products derived from "green" forests manufactured with "green" technologies. Conserving forest biodiversity has emerged as an integral part of "green" or sustainable forestry.
- While forests are located within the boundaries of sovereign nations, their ecological roles (such as carbon and hydrological cycles, habitats for migratory species, watersheds of international rivers) extend beyond national jurisdictions. The ecological role of forests at the trans-boundary and global levels has imposed extraterritorial pressures on nations to conserve, manage, and use their forests sustainably.
- Forests are the richest source of biodiversity on earth (see Chapter 1), and biodiversity is considered by some to be a "world heritage" – for the common benefit of all humankind. Conversely, nations with a rich source of biodiversity, particularly in tropical regions, claim "ownership" of the valuable genetic material and economic benefit that their resources possess.
- The concern to conserve and use biodiversity for present and future generations has raised ethical questions of intergenerational equity as well as the notion of shared responsibility.

Canadian Perspective
Forests, the most prominent feature of the Canadian landscape, cover about 50% of the country's land area. The value of shipments of Canadian forest products amounts to about $50 billion per annum. Canada is a major exporter of lumber, pulp, and paper, providing about 20-25% of the international trade in forest products. As a significant part of our country's economy, the Canadian forestry scene has witnessed the concerns that have evolved from local to global levels. The Canadian Forest Strategy, formulated in March 1992 and endorsed by the Canadian Council of Forest Ministers (CCFM), recognizes that the stewardship of Canada's forests is in our national socio-economic and environmental interest and is also part of maintaining a healthy global environment (CCFM 1992). While the word *biodiversity* was not even mentioned in a National Forest Strategy completed in 1988, the strategy formulated in 1992 explicitly recognizes the need to conserve Canadian forest biodiversity and proposes concrete actions and targets towards this objective.

We have seen – in response to the Brundtland Report to commitments made at the United Nations Conference on Environment and Development (UNCED), to green consumerism in the international marketplace, and to environmental concerns voiced by Canadians and others – a flurry of activities to define sustainable forestry in Canada (Maini 1989, 1993). A wide range of provincial, professional, environmental, and industrial organizations have proposed principles, criteria, guidelines, and indicators for sustainable forestry. A review of these proposals shows that of 33 documents analyzed, 26 recognize conservation of forest biodiversity as an important element of sustainable forestry (Buchanan 1993a, 1993b). The Canadian forestry community – including politicians, policy advisors, scientists, industry, and concerned citizens – are exposed, however, to as much fiction as fact and to slogans that lack scientific basis. At the same time, the Canadian forestry community is making serious commitments to sustainable forestry and is leapfrogging towards managing forests as ecosystems (Maini 1994).

Policy Environment
From the point of view of forming forest policy to conserve, monitor, and use forest biodiversity in Canada, the following elements should be considered:

• Despite the initial deforestation of about 15 million ha by European settlers (Maini and Carlisle 1974), the forest cover in Canada has not changed significantly during the past 50 or more years.
• Only 10% of Canada's forests are privately owned, 80% under provincial jurisdiction and 10% under federal jurisdiction. Provincial and federal governments have major roles and responsibilities with regard to Canada's forests.
• The number of species found in Canada's forests is estimated to be about 200,000 (Boyle 1991). Fewer than half have been described. According to Bunnell (1990), 76% of Canada's terrestrial mammals and 60% of Canada's breeding birds are forest dwelling.
• Many forest-dwelling species migrate to the United States, Mexico, and Central America. Canada's forest biodiversity policies must include transboundary and international considerations.
• Forest land and biodiversity are dynamic in both time and space. For example, some changes in the forest cover on a given land unit occur over decades to centuries. Some species are still migrating northward following deglaciation. The impact of anticipated climate change and depletion of the ozone layer on forest biodiversity has yet to be assessed (e.g., Hebda 1997).
• Perturbations such as fire, insects, and diseases are an integral part of boreal and temperate forest ecosystems. We must recognize the role of natural disturbance when we talk of managing and conserving biodiversity.

- The consequences of forest harvesting and subsequent regeneration for forest biodiversity have not been systematically researched.
- The Canadian forestry community is under pressure from various sources to practise environmentally sustainable forestry. The pressures include the interface between environment and trade; demands to maintain a healthy environment at the local, national, regional, and global levels; and ethical responsibility to future generations. These pressures are both internal (domestic) and external (international). Policies and practices aimed at sustainable management of forests must withstand the scrutiny of national and international political, policy, scientific, and environmental communities as well as that of consumer groups.
- Biodiversity has emerged as an issue of high priority on international political, policy, and scientific agendas. The worldwide political support for this issue, as evidenced by the signing of the Convention on Biological Diversity at the Earth Summit in Rio de Janeiro, outstripped the scientific knowledge of the issue. Biodiversity is a critical feature of the basic life support system, and its conservation is recognized as an important element of sustainable forestry (Maini 1989, 1993). Many organizations and countries are still searching for appropriate ways to meet the commitments they made under the Convention. We are experiencing uncertainty about what we know and what we need to know to formulate intelligent policies and practices. In view of our inadequate knowledge of the structure and function of forest biodiversity in Canada, initial policies on forest biodiversity should be based on the precautionary principle.
- Canada's forests are immense and cover about 420 million ha. Unlike Europe, in Canada the present and potential resources dedicated to research and to monitoring forest biodiversity are very limited. We will have to devise a uniquely Canadian approach that includes extensive management at the meso to macro levels.

Challenges Ahead

In formulating intelligent policies and practising environmentally sound programs to conserve, use, and monitor forest biodiversity, the scientific and policy communities face many challenges, including the following:

- We must learn to communicate to the policy-making and political communities the best available knowledge on biodiversity to serve as a basis for robust and flexible policies.
- In creating flexible policies, we must harmonize political, scientific, and environmental time horizons that range from years to decades to centuries.
- We must create robust policies and programs that also include long-term commitment to monitoring so that we can assess progress towards

attaining the stated policy objectives, feedback mechanisms, and mid-course corrective actions.

We face three challenges: communications, harmonization of time horizons, and monitoring policy. They deserve thoughtful consideration.

Communications

Different segments of society seek information organized in different ways. Initially, observations begin with data, which are transformed successively into statistics and information (such as scientific papers on individual issues). Scattered but related information is further transformed into knowledge (broad generalizations) through insightful analyses and reviews (not inventories of information; Figure 2.1 [top]). The policy community is generally interested in broad scientific generalizations about an issue (Figure 2.1 [bottom]) and is unable to cope with specialized and detailed scientific information or statistics (Figure 2.1 [top]). Senior science "statesmen and stateswomen" are needed to help translate scientific knowledge into policy-relevant terms.

The scientific community must recognize that different components of society are concerned with different levels of scientific information or knowledge (Figure 2.1 [bottom]). Whereas science technicians are concerned with micro-level change (such as changes of biota in experimental plots), scientists are concerned with system behaviour and are often fascinated by the noise in the system. On the other hand, policy analysts are generally concerned with broad trends, whereas policy advisors need to consider and address more complex issues, including cross-connections with other policy issues (for example, considering whether setting aside a given area as a national park would close lumber mills and cause unemployment, and whether ecotourism would compensate for the jobs lost through mill closure). Senior decision-makers and politicians are often concerned with "top of the pyramid" questions such as, "Are we winning or losing?"

Time Horizons

The policy process, involving long-term issues such as forest biodiversity, is further complicated by the fact that the political and scientific communities, as well as ecosystems, operate on different time scales. Changes in environmental systems (for example, in biodiversity, climate, forest ecosystems) involve decades and centuries, whereas the scientific and policy communities think in terms of 15 or more years and the political community usually thinks in terms of less than five years (Figure 2.2). How should the scientific community convince the political community of the need to operate under the precautionary principle to establish long-term monitoring systems and set aside appropriately sized (possibly extensive) landscapes as ecological

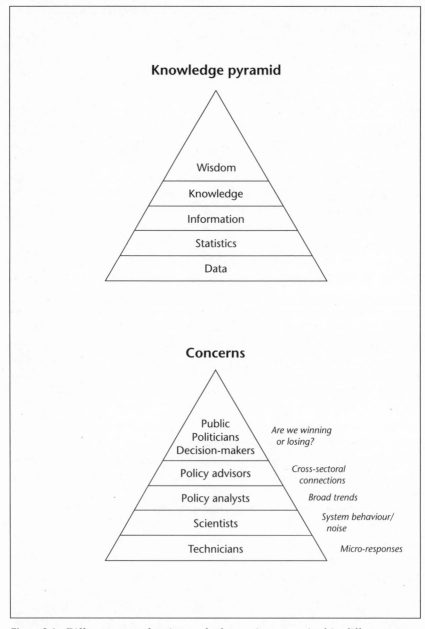

Figure 2.1 Different parts of society seek observations organized in different ways. (Top) Observations are successively transformed from data through statistics to knowledge and wisdom. (Bottom) The questions posed by different parts of society seek observations that are more or less well integrated (integration and synthesis increase towards the apex of the triangle).

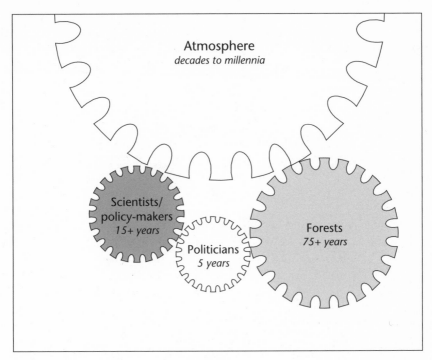

Figure 2.2 Different time scales of subsystems of a complex environmental issue such as forest and atmospheric change.

reserves, which would mean foregoing immediate, short-term economic and political benefits?

Monitoring and Mid-Term Policy Adjustments

Billions of dollars are being spent worldwide on a variety of environmental monitoring programs. Many are relatively new and their effectiveness is still unknown, particularly those concerned with biota. The monitoring of physical systems (such as the atmosphere or oceans) appears to be significantly more advanced than the monitoring of biological systems in terms of conceptual framework, policy relevance, and value for money. Biota monitoring must:

- have a sound scientific basis
- be diagnostic and help understand the state of a system
- permit an assessment of the stated policy objectives (Figure 2.3)
- include feedback to the policy process to enable mid-course corrections (Figure 2.3).

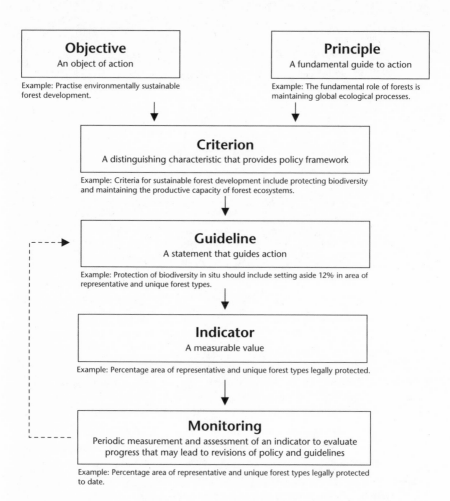

Figure 2.3 Definition, explanation, and hierarchical relationship of selected terms used in sustainable forest development (from Maini 1993). Policies aimed at attaining certain objectives must provide for systematic monitoring to enable mid-term corrective actions.

In the case of biodiversity monitoring, where the focus ranges from molecules and mounds to mountains, it is important that monitoring programs do not focus only on the so-called "keystone species."

Approaches to Formulating Policy on Forest Biodiversity
In formulating approaches to conserve and manage Canada's biodiversity, a wide range of factors should be considered. These include the extensive area of forests in Canada; our limited knowledge of forest biodiversity; the

limited availability of technical, scientific, and financial resources; the dynamic nature of forest ecosystems in the short, medium, and long term; local, national, trans-boundary, and global dimensions; the role and consequences of natural and manmade disturbances on forest biodiversity; and the economic benefits derived from Canada's forests. Consequently, policies must:

- apply the "precautionary principle"
- be experimental
- be robust in the face of political whims
- employ the best available scientific knowledge
- provide for monitoring so that we can progress towards attaining stated objectives and revise policies as new information becomes available.

Ludwig et al. (1993) offered interesting and damning observations on resource exploitation and conservation of fisheries within the overall policy environment of uncertainty. Unfortunately, the fisheries sector in various parts of the world has not based its harvesting on the precautionary principle. The forest sector differs from the fisheries sector in at least two respects: (1) the consequences of forest policies and practices are highly visible, and (2) the marketplace is demanding green products from green, "biodiverse" forests.

Environmental concerns expressed by Canadians and green market demands are likely to have significant influence on Canada's approach to conserving and managing forest biodiversity. The complexities of dealing with the large biota, however, remain. Reid et al. (1993) proposed a number of biodiversity indicators for policy-makers. Bunnell (1997, 1998) discussed the major implications of what the public wants in terms of sustaining biodiversity and offered suggestions on the essence of sustainable forestry. The most complete treatment of sustainable forestry, including a detailed approach to monitoring success, is that of the Scientific Panel for Sustainable Forest Practices in Clayoquot Sound, also known as the Clayoquot Scientific Panel (Scientific Panel 1995). The panel also emphasized the importance of the precautionary principle and noted that both management and policy should be adaptive.

At present, the Canadian policy communities engaged in formulating Canadian forest biodiversity strategy are faced with the following questions:

- How much natural forest should be protected as national parks, ecological reserves, and biosphere reserves, and where?
- What is the negative impact of forestry practices on biodiversity, and how can we minimize this impact?
- Are planted forests unacceptable in terms of conserving biodiversity?

- What is the impact of natural and artificial regeneration on forest biodiversity?
- What is the role of natural disturbances on maintaining forest biodiversity?
- What is the role of seed banks and gene banks in Canada's forest biodiversity strategy?

The Canadian approach should be extensive (large scale) rather than intensive, recognizing the environmental as well as economic benefits of forests and accommodating natural, modified, and planted forests. The approach should include local, national, trans-boundary, and global dimensions. The Canadian forest biodiversity policy should include both in situ and ex situ conservation actions. It should also include a well-considered diagnostic monitoring program that is linked with the policy process so that we can revise policies, regulations, and practices.

Literature Cited

Boyle, T.J.B. 1991. Biodiversity of Canadian forests: Current status and future challenges. Forestry Chronicle 68:444-52.

Buchanan, K. 1993a. A comparison of forestry principles proposed by several Canadian institutions with those of the National Round Table on Environment and Economy. Canadian Forest Service. Unpublished report.

–. 1993b. A comparison of forestry principles as proposed by several Canadian institutions with criteria as proposed by Dr. J.S. Maini. Canadian Forest Service. Unpublished report.

Bunnell, F.L. 1990. Biodiversity: What, where, why, and how. Pp. 29-45 *in* Wildlife Forestry Symposium: A workshop on resource integration for wildlife and forest managers. Symposium Proceedings, 7-8 March 1990, Prince George, BC.

–. 1997. Operational criteria for sustainable forestry: Focusing on the essence. Forestry Chronicle 73(6):679-84.

–. 1998. Overcoming paralysis by complexity when establishing operational goals for biodiversity. Journal of Sustainable Forestry 7(3/4):145-64.

Canadian Council of Forest Ministers (CCFM). 1992. Sustainable forests: A Canadian commitment. CCFM report.

Hebda, R.J. 1997. Impact of climate change on biogeoclimatic zones of British Columbia and Yukon. Pp. 13.1-13.15 *in* E. Taylor and B. Taylor (eds.). Responding to global climate change in British Columbia and Yukon. Volume 1 of the Canada Country Study: Climate impacts and adaptation. Environment Canada and BC Ministry of Environment, Lands and Parks, Vancouver, BC.

Ludwig, D., R. Hilborn, and C. Walters. 1993. Uncertainty, resource exploitation and conservation: Lessons from history. Science 260:17-36.

Maini, J.S. 1989. Sustainable development and the Canadian forest sector. Discussion paper presented to the Canadian Council of Forest Ministers, 6 October 1989, Niagara Falls, ON. Forestry Canada, Ottawa, ON.

–. 1993. Sustainable development of forests: A systematic approach to defining criteria, guidelines and indicators. Conference on Security and Cooperation in Europe, Seminar of Experts on Criteria and Indicators of Sustainable Forestry, September 1993, Montreal, PQ.

–. 1994. Canada's commitment to sustainable forestry. Canadian Pulp and Paper Association, Wood Pulp Section Open Forum, Montreal, PQ.

Maini, J.S., and A. Carlisle. 1974. Conservation in Canada. *In* J.S. Maini and A. Carlisle (eds.). Conservation in Canada: A conspectus. Environment Canada Publication Series No. 1340, Ottawa, ON.

Reid, W.V., J.A. McNeely, D.B. Tunstall, D.A. Bryant, and M. Winograd. 1993. Biodiversity indicators for policy-makers. World Resources Institute/International Union for Conservation of Nature and Natural Resources, Gland, Switzerland.
Scientific Panel for Sustainable Forest Practices in Clayoquot Sound. 1995. Report 5, Sustainable ecosystem management in Clayoquot Sound: Planning and practices. Ministry of Environment, Lands and Parks, Victoria, BC.

3
Genetic Diversity for Forest Policy and Management
Gene Namkoong

Introduction: Simplicity Is Not Enough

Concerns for biodiversity reflect perceived threats to many organisms at different taxonomic, temporal, and geographic scales. While I believe that substantial threats exist, I also believe that unless we analyze the dynamics of biodiversity, oversimplified solutions will become the only alternatives for forest management. An unexamined general assumption is that there are relatively few ultimate causes, that these can be politically solved, and that strong and simple functional relationships exist among all organisms and among all scales of concern such that interventions cause unwelcome and pervasive destruction. If this assumption were true, land managers would need to know little about the structure or dynamics of diversity because single actions would have predictable effects. With the perception of a broad threat and the suspicion of unified causes and effects, people naturally hope that simple solutions, such as withdrawing all human intervention or segregating land uses into separable compartments, will solve the problem.

The simple solution of segregating land-use zones, however, generates problems. Stands or compartments are often not effective barriers; boundary effects depend on interacting spatial and temporal patterns in a mosaic of neighbourhoods. Colonization by individuals or groups and gametic migrations are influenced by distance and vectors that may span stand boundaries and affect the internal dynamics of populations within neighbourhoods (Young et al. 1996). Furthermore, many solutions for conservation of biodiversity designate the vast majority of lands as "multiple-use" or "multiple-function" zones within which mixed management objectives are pursued by multiple methods.

While the threat may be broad and the biota are indeed interconnected, simple solutions that either place trust in manipulative methods or forego choices in human interventions can be counterproductive if uniformly applied. Humans, having been a part of the functioning biota for many millennia, have a clear claim for moral consideration and are now faced with choices that require our best scientifically informed guidance. Foresters have traditionally developed management plans at the stand or compartment level, with operations limited to planting, thinning, and harvesting. But management now includes the genetic structure of planted and resident species as well as ecosystem management at broad scales, and these are the same in neither concept nor execution. No single approach to understanding or management, whether at the genetic or ecosystem level of biological organization, can be sufficient. In this chapter, I focus on the genetic dynamics of the expansive and contractive forces that affect genetic conservation. From this genetic perspective, I hope to contribute to an integrated concept of forest management. Because of concerns about artificial regeneration constraining genetic diversity in forest plantations, many of my comments are specifically about trees. The principles discussed are more broadly applicable, although some techniques would be more difficult to apply to freely moving organisms.

Genes Are Not Enough

Genetic diversity is a necessary but insufficient condition for the survival of most species in the short run and for all species in the long run (see also Chapter 4). Because interactions among different species affect competitive and mutualistic interactions and their heritable components, they also constitute selective "environmental" effects. Species also affect pollen and seed movement and mating patterns of other species, and therefore also help determine the distributional patterns of species, populations, and migrations.

These heritable patterns of variation and their continuing evolution are important features of what we mean when we say we wish to conserve this or that species of animal or plant, and when we say that conservation goals include conserving evolutionary potential (Frankel 1974). The concept of evolution thus changes our idea of what we wish to conserve from one of pre-Darwinian typology to one of dynamic conservation (Eriksson et al. 1993); it forces us to consider not only that species diversity depends on the genetic resources of species but also that genetic diversity depends on the diversity of species interactions. *Therefore genetic and species diversity are not independent.* The goals and methods of managing genetic diversity differ, however, from the goals of managing for species diversity in communities and from those of managing community diversity in forests. Each biotic level has unique elements and dynamics that are appropriately managed at those levels, and, while recognizing that inter-level interactions are part of their dynamics (e.g., Levin 1992; Bunnell and Huggard 1998), each biotic level is distinct and requires different methods of management.

At the genetic level, because genes themselves undergo change and evolution, because new multi-locus genotypes are continually being created and lost, and because populations change in size and genetic composition, we can influence, by design or accident, the forces that affect the size and structure of the genetic endowment. This observation implies that any attempt, as a single conservation strategy, to save all alleles or all allelic combinations in reserves or static collections is not only physically impossible but could be detrimental to a dynamic system. *The goals of gene management should therefore be directed towards managing dynamic rather than static systems.*

Genes are the finest level of biological concern, but conserving individual genes is seldom seen as the purpose of genetic management. While genes are self-replicating, they have no consciousness or capacity for independent function, and do not carry the same values as may be inherent in individuals or groups of individuals. It is therefore difficult to accept, as an end, an obligation to conserve one gene or a collection of genes. Genes are necessary elements that serve larger goods, such as species survival, that in themselves can carry inherent values and obligations for conservation. Moreover, because genetic diversity is a group property, the duties for its conservation lie with the group, population, stand, or larger level of organization rather than with the individual organism.

Because populations and species are persistently in a dynamic evolutionary state, they change in the composition of their elements. The current levels and structures of their variations are passing features of their evolutionary processes, and thus cannot serve as normative goals for management policies. The present state of diversity is clear evidence of one evolutionarily feasible state, but it is not necessarily the "best" or only possible state that could have resulted from even the same basic forces. The present state may well serve as a comfortable model but, Dr. Pangloss[1] notwithstanding, programs for maintaining or re-establishing any past or present genetic state would not necessarily serve a deeper good. The ethical challenges of defining values for future states of the biota are substantial, but the present state of the genetic distribution cannot be used to define appropriate targets. Biologists are therefore largely limited to providing the means by which multiple alternative futures may be attained.

Gene Management Options
Gene management options are varied and can be intensive, as in rescue programs for the Speke's gazelle (Templeton and Read 1983) or in crop breeding, or they can be non-intrusive, as in extending protection zones for critical populations or merely monitoring effective population sizes in reserves. They can be both, as when reintroductions with replacement genotypes are needed from ex situ sources for forest underplanting or supplementing

---

1   In *Candide,* Dr. Pangloss is Voltaire's caricature of the optimistic belief that "all is for the best in this best of all possible worlds."

natural regeneration, or when reintroductions are needed from zoos to the wild. Because both in situ and ex situ techniques of gene management can be used, we need not depend on land management techniques to control diversity. The questions addressed here include ecological goals for gene management, appropriate techniques, and appropriate measures in a forest management system that integrates ecological and genetic relations.

Goals

Gene management methods are largely limited to operations within the bounds of species. While some uncertainty exists about the limits of cross-ability among species, notwithstanding transgenic technologies, we will be limited for the foreseeable future to managing variation within species. Species differ substantially in their particular evolutionary history, in the effects that the various forces of evolution have on them, and in their capacity to respond to the forces of evolution and our capacity to influence those forces. All species, however, follow the same basic principles of evolution. They all respond primarily to the contributions of selection, constraints on population size, and the sharing of genes among populations through migration, and all ultimately rely on mutation as a source of new variants. Species carry mutational, inbreeding, and outbreeding loads, and they lose and gain alleles as whole species and in local populations. Species are thus a primary unit of the goal of conserving biological diversity.

For any species that may have existed as a single, large, random-mating population in stable equilibrium, genetic variation can be maintained most easily by ensuring that certain minimum effective population sizes persist. In the face of random losses caused by demographic stochasticity, by environmental changes within certain bounds of variance, and by large-scale random catastrophes, we can derive minimum population sizes having low probabilities of extinction (Soulé 1987). To maintain enough genetic resources to ensure that population sizes are sufficiently large to avoid inbreeding depression, it is adequate that a few tens of mature, reproductively capable individuals intercross every generation. For any species, a few hundred independent genotypes are needed to maintain quantitative genetic variance at nearly the same level as in an infinitely large population, and a few thousand are needed to ensure high probabilities that many rare alleles will remain in a population (Namkoong 1988).

These concepts have been developed elsewhere. The major implication to be drawn for managing land areas is that genetic considerations simply increase the number and variety of factors that should be measured in conservation programs. Because the main goal of gene management is to protect the evolutionary potential of species, however, and because the current state of the system is an initial condition and not an end goal, it is the size and pattern of variation that should concern management.

Techniques

Genetic management by accumulating large random collections is one technique, but not the only one available. Minimum numbers can be used as guides for random processes and for random collections of genes and individuals. For many purposes, including that of maintaining evolutionary potential, targeted selection is more effective than large, random collections. If variables are normally distributed, accumulating random entries has a low probability of increasing the limits of the range of response or the range of conditions over which individuals can survive. Increasing the genetic variance in traits of adaptive significance is better attained by maintaining self-regulating populations over a wide environmental range than by merely accumulating random entries.

Targeting selections for diversity at the individual and population levels, or selecting for developmental homeostasis at both levels by structuring populations, can decrease extinction probabilities. Anything that either restricts migration or increases exchange among largely distinct populations can encourage rapid change in genetic composition. Thus such simple techniques as targeting samples to populational outliers or to environmental or distributional extrema will increase diversity with or without divergent selection. Designating such populations for in situ conservation can be more efficient for capturing diversity than saving single larger populations. When threats can be anticipated, divergent selection to favour different behaviours or phenotypes, by breeding or by selecting seed parents in naturally regenerated forests, can be a powerful tool for increasing variance. These applications of mating and selection control can be more or less intensive than the species may have recently experienced, but for many species, they are not qualitatively different forms of evolution in genetic diversity.

Selection in segregated populations can rapidly change the amount and structure of genetic variation and, while losing some alleles in some populations, may gain alleles in others. Long-term selection experiments have also clearly demonstrated that new mutations and divergent selection can rapidly increase total genetic variance (MacKay et al. 1994). The combination of directing selection and erecting mating barriers can produce rapid evolutionary change and is a more effective means of sustaining diversity than accumulating large, random collections. Total genetic variances can, in fact, be increased. In addition, as long as populations are at sufficient size that they remain viable, losses of genetic variance within populations can be negligible and new mutant alleles can attain higher probabilities of being saved than would occur without intervention. The sizes of such "guided" samples need not be large to maintain most of the genetic variance that would be maintained in naturally occurring populations many times their size (Brown 1989). Thus, directed or targeted sampling of extant populations can be more effective in ensuring high probabilities of saving alleles than can random collections, and directed selection can be more effective in saving mutant alleles than can random accumulations. This approach can be managed by selecting stands in different environments or by limiting regeneration to parents exhibiting different behaviours in the different stands.

The probability of rare but unknown (and potentially useful) alleles occurring in a targeted sampling and selection system is not less than in a single population of the same total size. It is not clear, however, that the probability of recognizing such rare alleles is at all high, or that the function of such an allele may not otherwise be mimicked by other alleles. The contributions of such alleles in the world's seed banks has not been great; in terms of capturing and making use of exotic alleles, several smaller core collections selected for genetic diversity are more efficient than large random collections (Brown 1989). Maintaining large populations may be desirable for many reasons, but to efficiently maintain genetic variance is not one of them.

Appropriate Measures
For genetic resources, several levels of structure may exist, and diversity at each level can be subject to different management techniques. Among species, genera, and higher levels of genomic organization, it is possible to compose a diversity measure based on genetic or taxonomic distances (Harper and Hawksworth 1994). These distances may be measured in several ways, and by selecting more genetically distant taxa, the genetic diversity at those levels can be maximized. Such measures as species richness can be extended to genus or family, and targets for inclusion in the conservation pool can be identified for their probable contributions of unique genes or chromosomes, or large selection of genomic diversity. Within species, similar and multiple levels of diversity among populational, provenance, or other hierarchies also exist. A commonly used measure of genetic distance is Nei's, but several others are described by Gregorius and Roberds (1986). Hierarchies of distance statistics can also be derived for multiple loci (Weir 1990).

Most species in fact are subdivided to some extent, having different alleles and different frequencies of alleles among subpopulations. Over large ranges, such subdivisions may be reinforced by mating barriers and by environmental differences that modify selection. These mating barriers and environmental differences may be important features by which species, through reaction to environmental changes, maintain their evolutionary capacities. A major concern of conservation biology is the detrimental consequence of population subdivision and habitat fragmentation in reducing the effective population size, which may induce inbreeding depression, reduce levels of genetic variance, and increase the probability of losing low-frequency alleles. The concern is well taken for species with few populations, most of which are small and endangered in themselves (Soulé 1987). As discussed elsewhere (see Chapters 4 and 8), fragmentation can generate many hidden dangers, and the probability of local extinctions can be much higher than one might expect. In addition to viability, reductions in effective population size can be large if mating frequencies and contributions to the progeny pool generated by different fecundities are highly unequal, and generational fluctuations in mating pool size and iteroparity[2] can reduce the independence of genes in populations (see review by Hartl and Clark [1989]).

If short-term survival can be otherwise ensured, however, the creative effects of selection among population subdivisions can generate useful interpopulational diversity (Eriksson et al. 1993). Even for alleles that have no known distributional characteristics and for those that are rare and widely dispersed, population subdivision neither increases nor decreases the probability of conservation. In the case of alleles that have different frequencies or that are present in some populations but not in others, it is possible to increase the probabilities of maintaining them by targeting selection among populations for divergence and by ensuring the survival of outlier populations (Namkoong 1988) where unique or private alleles (Slatkin and Maruyama 1975) may exist.

For species that merit intensive gene management, the present levels of population subdivision can serve as a reasonable starting point for further diversification of populations. In maize (Goodman 1985) and in trees (Eriksson et al. 1993), naturally occurring population differences have proven useful. We should note that if populations undergo changes in allelic frequency during cyclical events such as those that occur over successional stages, the allelic frequencies can never be expected to stabilize at equilibrium frequencies and would rarely be the same in different populations for selective as well as stochastic reasons. Hence multiple population sampling and multiple population maintenance may be important not only as tactics for gene conservation but also because they are significant features of the patterns of variation of the species. Sampling among different portions of the species may be more than a useful way to carry out gene conservation; it may be an important feature of those species we wish to conserve. Therefore the genetic variation, the population sizes and breeding system that can maintain such variation, and the structures of variation and their rates of response to divergent selection are all candidates for features of evolution that we might want to use as measures for forest management.

It is also true that most species have many genes and, while most species obey the same rules of evolution, each is subject to different pressures of

---

2 Iteroparous species have more than one breeding period; semelparous species, such as salmon, are restricted to one.

mutation and selection, and hence also of drift. Forest tree species seem to have among the highest levels of variation at isozyme loci, and also have high levels of variability in quantitative genes (Namkoong et al. 1988). This leads to the expectation that many as yet unexplored species could perhaps have significant allelic variations in a majority of their loci. There might be little problem with this if all alleles shared the same pattern of variation and all were equally significant for conservation. If that were the case, then we could target the rarest of alleles that might be significant; if we could ensure a reasonable probability of their survival, we would expect that all other alleles would also be conserved. Unfortunately, variation patterns for isozyme alleles believed to be selectively neutral are often not congruent with patterns of heritable, adaptively significant traits. It is therefore desirable to use several measures of genetic variation to target conservation efforts.

Components of Genetic Management
While all species obey the same evolutionary rules, they are all different in genetic variation, population sizes and subdivisions, selective history and influences, and mating systems – and hence in their resilience, stability, capacity to respond to challenges, endangerment, and capacity to be influenced by us in their survival and future evolution. We can monitor and manage genetics by various means, ranging from intensive breeding and genetic rescue to merely ensuring that large reserves exist where certain species occur. For the most intensive techniques, we can use even single individuals in a hybrid breeding program to ensure that at least one chromosome sample is viable, recovering a species or genome in later segregating generations. Intensive breeding for the capacity to endure inbreeding is also possible (Templeton and Read 1983).

Both of these techniques do not require maintaining large numbers because they are used when few alleles exist. For random alleles and for maintaining genetic variation in quantity, large numbers of individuals ensure that rare alleles have reasonable probabilities of existence. Relevant calculations of appropriate numbers are presented elsewhere (e.g., Namkoong 1988). The numbers are not large if mating can be controlled, but if mating cannot be controlled, the need for redundancy increases. Thus with high degrees of knowledge and control, we need relatively small populations or sets of populations to ensure gene conservation, and more can be done to increase variation. The problem is that these conditions exist for a very small minority of species. In the genetic management of forest tree species, breeding can strongly influence the use and maintenance of genetic variation in such a way that growth and form are easily changed without decreasing isozyme variation (El-Kassaby and Namkoong 1994), but tree breeding is limited to a few dozen species out of the tens of thousands of woody species that compose the dominant vegetation in the world's forests. Similarly, for herbaceous vegetation, about 0.1% of all vascular plant species are included in gene conservation programs by the International Plant Genetic Resources Institute; the percentage for large mammals is only slightly better. A significant management question is how to develop a system to set priorities among species and how to allocate effort among them.

The state of genetic management and information can be easily enhanced with presently available techniques, and that is a clear priority (e.g., National Research Council 1991). Even increasing the number of included species by an order of magnitude, however, would still leave a vast majority of species that cannot be intensively managed. For genetic management, efforts might be reasonably allocated on the basis of three criteria: recognized value, vulnerability or need for genetic management, and likelihood that genetic management will be effective. Value depends on considerations such as those noted elsewhere in this book (Chapter 8, for example), and includes both economic and ecological significance. Need and vulnerability can be estimated using population size, potential viability, and the potential for disrupting reproduction systems. The array of genetic management techniques could then be allocated based on relative payoffs for the alternative methods and species. Genetic management would presumably be marginally less costly if integrated with other land management tactics, but ex situ techniques that separate genetic and land management need not be expensive. Intensive management and heroic rescue methods such as those conducted for Speke's gazelle and the California Condor will not often be affordable and may detract inordinately from other types of effort.

It would be easiest for a forest management plan to focus activities around a single mode of action such that if one level of management were successful, success at all other levels of the biota would follow. It does not appear likely that gene management can do this, nor is ecosystem management or any other single level of management likely to fill the role (e.g., Bunnell and Huggard 1998). Nevertheless, it should be feasible to develop a management system around programs that can simultaneously satisfy multiple objectives, and to supplement programs initially designed at one biotic level with efforts aimed at the elements of the genetic or other subsystems that are most endangered or least conserved by the initial system. From this perspective, most species would have to rely on ecosystem- and landscape-level management for their genetic conservation; only species incapable of self-sustaining evolution would require intervention. For most species, management would consist of establishing large and possibly several reserves or areas of specific forest age classes in managed forests. Where contiguous areas are fragmented, corridors may be needed. This is essentially the approach taken by Frankel (1983) and in British Columbia (the combination of a Protected Areas Strategy and the Forest Practices Code). While the goals of genetic and ecosystem management are not the same, conserving multiple samples of multiple ecosystems in diverse habitats might be appropriate, resulting in a "base collection" of genes for many species. In this sense, a subcollection of ecosystems might be designated as a "core collection," in the same sense that Brown's (1989) core seed collections serve for a single plant species.

This system would require inventories for at least sampled subsets of classes of species present in reserves, parks, or other forms of protected areas, as well as projections of their evolutionary capacity in terms of population size, number and distribution of populations, and environmental distribution. Thus various classes of organisms would require sampling, and special effort would be targeted towards those classes that, because of their special viability or special reproductive systems, were more vulnerable to loss than might be expected from their numbers alone (for example, terrestrial breeding salamanders [Bunnell et al. 1997]). Some classes of species and some species (possibly wide-ranging carnivores) will not be readily maintained in these programs. If they are deemed of sufficient value and likely to benefit from intensive genetic management, they will have to compete with other species for funding of genetic management.

An alternative system for setting priorities is to identify species having such high economic or ecological significance that if they are conserved, ecosystems and economic viability are also likely to be conserved (e.g., Walker 1995). The genetic structure and variability of these species would then be targeted for monitoring and development, and the principal difficulty would be to include enough of these indicator and keystone species that they would, in fact, serve to support all necessary ecosystem functions. By this method, rare species would have no priority unless they served an economic or ecological function, while targeted species could include obscure indicators of functionally important but sparsely distributed processes. This is the approach taken by Schoenwald-Cox (1983) and the Scientific Panel for Sustainable Forest Practices in Clayoquot Sound (the Clayoquot Scientific Panel) (Scientific Panel 1995); it can be considered complementary to that of Frankel (1983). Ultimately, all species would be included under the categories of immediate and well-known significance, of secondary and potential value, significant for the well-being of the species of primary interest, and so on. Such a system of species priorities, advocated by Norton (1987), would have to be augmented by the degree of intervention and the kinds of genetic variation that would be of greatest significance to species in each category.

I suggest that either of these approaches would provide a reasonable framework for gene conservation and that macro and micro time and geographic scales can define the time and size scales necessary for defining the structure and levels of genetic variation required for species or landscape levels of conservation. In those species chosen for genetic survey and management, we should survey the levels of genetic variation of both neutral genes and selectively significant genes, and estimate the structure of that variation and the effective population sizes. For species with small effective population sizes or with restricted mating populations and low levels of extant genetic variation, we could make a special effort to ensure continued viability and the recovery of genetic variation. Some endangered species may merit neither the effort required for genetic rescue nor the management efforts required to increase their probability of survival. I suggest that intervention efforts should require not only that species have a utility value and that their genetic status be endangered but also that the means of genetic management be such that we can expect effective genetic rescue. If we must choose to allow extinctions, then gene conservation would give rescue priority to species that contribute most to allelic diversity.

Figure 3.1 shows the different genetic management methods. Direct management of genetic variation is emphasized at the top of the chart, while land management is emphasized at the bottom. All forms of management affect the size and distribution of the genetic variation, and diversity among multiple populations is useful in each form of management. Although genetic variation among and within populations is subject to increases and

Breeding
Screening, testing, and evaluation
Collecting
Sampling
Survey
Managed reserves
Strict nature reserves

Figure 3.1   Genetic management methods in descending order of management intensity (from direct genetic sampling to dependence on random collections).

decreases, the less direct forms of management require larger population sizes to ensure conservation through undirected redundancy.

Management methods can also be ordered against the different goals of genetic management (Figure 3.2). Goals may be grouped into four sets: (1) species with critical economic or ecological significance; (2) species with potential significance as economic or ecological substitutes or having secondary

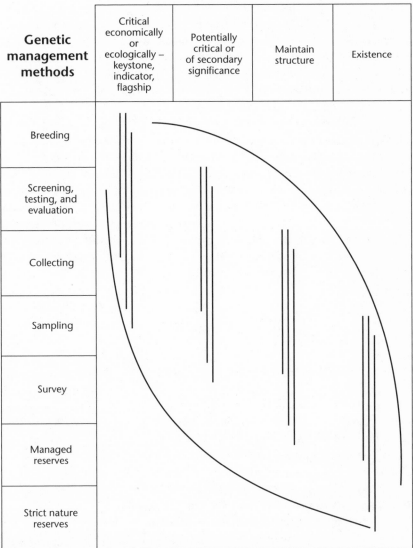

Figure 3.2 Matrix of genetic management methods and values. Management methods are not all equally appropriate for different goals of management.

and supportive functions; (3) species that are significant and known, and for which it is useful to maintain diverse genetic structures; and (4) species whose significance is unknown and whose value may lie only in their existence. The appropriate forms of genetic management for each of these classes lie along a diagonal extending from the upper left, where high specific value justifies intensive management, to the lower right, where extensive land management accommodates species of existence value only.

A strategy for conservation suggests itself. The first step is to ensure that programs already in place include the vast majority of species we can sample. This involves monitoring samples of areas and taxa under different intensities of land use. At present it would include a few tens of species in the upper left of Figure 3.2 (those with critical economic or ecological significance) and should include several tens of species in the second column (those with potential significance). The vast majority of species will fall into the systems of low-intensity management.

The second step is to look for the exceptions to the general program. In particular, attention should be given to species that would ordinarily lie in the lower right of Figure 3.2 but whose population sizes, distributions, or environmental sensitivity threaten their viability. These species consist of sparsely distributed individuals that exist in few populations, that are specialists, and that have low birth rates and vulnerable reproductive systems. Even if they are of low economic or ecological significance, they may be salvable, and allocating conservation efforts to them should be considered.

The question of measuring genetic diversity can now be put in terms of the management approaches shown in Figure 3.1. For management approaches using genetic variation for intensive breeding or development, direct measures of genetic variation in significant traits is of primary interest. Of secondary interest are measures of isozyme variation or other presumably neutral DNA variations. For both types of measures, the distribution of genetic variation helps to determine whether initial sampling is sufficient and whether future evolution of new structures is feasible or useful.

For approaches involving less direct genetic management and more reliance on large redundancies and random gene collections, population size estimates and the levels of effective population sizes could be used as surrogates for direct estimates of genetic diversity. To identify species that should be managed more intensively or for which either an escalation or de-escalation of intensity is warranted, sampling neutral alleles is useful for indicating potential problems. Both the number of populations and the number of individuals within populations would be useful parameters.

No single measure and no single set of genes can serve all purposes for genetic diversity, but with some forethought, a finite number of measures can be used to answer most of the genetic resource questions affected by a broad definition of forest management.

Literature Cited

Brown, A.H.D. 1989. The case for core collections. Pp. 136-56 *in* A.H.D. Brown, O.H. Frankel, D.R. Marshall, and J.T. Williams (eds.). The use of plant genetic resources. Cambridge University Press, Cambridge, UK.

Bunnell, F.L., and D.J. Huggard. 1998. Biodiversity across spatial and temporal scales: Problems and opportunities. Forest Ecology and Management. In press.

Bunnell, F.L., L.L. Kremsater, and R.W. Wells. 1997. Likely consequences of forest management on terrestrial, forest-dwelling vertebrates in Oregon. Oregon Forest Resources Institute, Portland, OR.

El-Kassaby, Y., and G. Namkoong. 1994. Impact of forest management practices on genetic diversity and its conservation. Pp. 205-13 *in* R.M. Drysdale et al. (eds.). Proceedings of the International Symposium on Genetic Conservation and Production of Tropical Forest Tree Seed. ASEAN (Association of Southeast Asian Nations)-Canada, Forest Tree Seed Centre Project, Chiang Mai, Thailand.

Eriksson, G., G. Namkoong, and J.H. Roberds. 1993. Dynamic gene conservation for uncertain futures. Forest Ecology and Management 62:15-37.

Frankel, O.H. 1974. Genetic conservation: Our evolutionary responsibility. Genetics 78:53-65.

–. 1983. The place of management in conservation. Pp. 1-14 *in* C.M. Schoenwald-Cox, S.M. Chambers, B. Mcbryde, and L. Thomas (eds.). Genetics and conservation. Benjamin/Cummings, Menlo Park, CA.

Goodman, M.M. 1985. Exotic maize germplasm: Status, prospects, remedies. Iowa State Journal of Research 59:497-527.

Gregorius, H.R., and J.H. Roberds. 1986. Measurement of genetical differentiation among subpopulations. Theoretical and Applied Genetics 71:826-34.

Harper, J.L., and D.L. Hawksworth. 1994. Biodiversity: Measurement and estimation. Philosophical Transactions, Royal Society, London 345:5-12.

Hartl, D.L., and A.G. Clark. 1989. Principles of population genetics. 2nd ed. Sinauer Associates, Sunderland, MA.

Levin, S.A. 1992. The problem of pattern and scale in ecology. Ecology 73:1943-67.

MacKay, T.F.C., J.D. Fry, R.F. Lyman, and S.V. Nazhdin. 1994. Polygenic mutation in *Drosophila melanogaster:* Estimates from response to selection of inbred strains. Genetics 136:937-51.

Namkoong, G. 1988. Population genetics and the dynamics of conservation. Pp. 161-81 *in* L. Knutson and A.K. Stoner (eds.). Biotic diversity and germplasm preservation, global imperatives. Kluwer Academic Publishers, Boston, MA.

Namkoong, G., H.C. Kang, and J.S. Brouard. 1988. Tree breeding: Principles and strategies. Monographs on Theoretical and Applied Genetics II. Springer Verlag, New York, NY.

National Research Council. 1991. Managing global genetic resources: Forest trees. National Academy Press, Washington, DC.

Norton, B.G. 1987. Why preserve natural variety? Princeton University Press, Princeton, NJ.

Schoenwald-Cox, C.M. 1983. Conclusions: Guidelines to management: A beginning. Pp. 414-45 *in* C.M. Schoenwald-Cox, S.M. Chambers, B. Mcbryde, and L. Thomas (eds.). Genetics and conservation. Benjamin/Cummings, Menlo Park, CA.

Scientific Panel for Sustainable Forest Practices in Clayoquot Sound. 1995. Report 5, Sustainable ecosystem management in Clayoquot Sound: Planning and practices. Ministry of Environment, Lands and Parks, Victoria, BC.

Slatkin, M., and T. Maruyama. 1975. The influence of gene flow on genetic distance. American Naturalist 109:597-601.

Soulé, M. (ed.). 1987. Viable populations for conservation. Cambridge University Press, Cambridge, UK.

Templeton, A.R., and B. Read. 1983. The elimination of inbreeding depression in a captive herd of Speke's gazelle. Pp. 241-61 *in* C.M. Schoenwald-Cox, S.M. Chambers, B. Macbryde, and L. Thomas (eds.). Genetics and conservation. Benjamin/Cummings, Menlo Park, CA.

Walker, B.H. 1995. Conserving biological diversity through ecosystem resilience. Conservation Biology 9:747-52.

Weir, B.S. 1990. Genetic data analysis. Sinauer Associates, Sunderland, MA.

Young, A., T. Boyle, and T. Brown. 1996. The population in genetic consequences of habitat fragmentation for plants. TREE (Trends in Evolution and Ecology) 11:413-18.

4
Biodiversity at the Population Level: A Vital Paradox
Gray Merriam

Introduction

Biodiversity at the population level is a paradox. At the population level, only genetic variation can be called biodiversity. Yet changes in diversity, no matter what the cause, always take place at the population level when mortality exceeds population recruitment. This observation is rarely clear in the confusion surrounding the subject of biological diversity.

Despite early scientific contributions (e.g., Wilson and Peters 1988), biodiversity was soon transformed from an ecological or evolutionary variable into a campaign slogan. Although many people still refer to its predecessors – alpha, beta, and gamma diversities (Whittaker 1972) – biodiversity has become, since the 1992 United Nations Conference on Environment and Development in Rio de Janeiro, an environmental fashion statement. It is a politically correct term for the outdated "balance of nature." Efforts such as the Global Biodiversity Strategy (WRI/IUCN/UNEP 1992) have tried to preserve opportunities for scientific input to this topic, but scientists themselves have different views.

Sagoff (1993) has argued, from philosophical roots, that ecologists are unlikely to approach biodiversity consistently because of basic differences in their philosophy of science. The group seeking predictive power has philosophical roots in Plato's ideals, using broad but simple concepts of hierarchies, steady-state analyses, or pulsed stability. Another group, with less interest in grand theory and mathematical laws, has philosophical roots in Aristotle and delights in nature's tiniest details. Anyone looking to science for answers about biodiversity must somehow untangle the political forces from the science, and must then be aware that values placed on biodiversity by two broad views of ecology serve to complicate perspectives. Platonic rationalists following systems ecology in search of grand theoretical predictions are no less zealous than modern-day naturalists using empirical details to value nature culturally and aesthetically.

Management issues have spun off from the use of the term *biodiversity* in policy statements (with little content), laws, and, indeed, a global UN convention. Such inverted and process-driven development of policy from politically appealing buzzwords causes crippling gaps between the knowledge produced by research and practical uses of that knowledge. The dangers for managers are first in believing that there really is some knowledge about this buzzword, and second in failing to achieve the imagined benefits ascribed to the concept. The danger for researchers is in falling prey to bureaucratic expectations of "science to fill the policy" instead of "science to critically assess the policy." Scientists judge statements by their ability to withstand refutation. When terms are vague, their challenge is not to devise meanings for the terms but to relate the terms to issues that are better understood.

Although much of the biodiversity discussion is about species loss, this phenomenon is composed of population changes and, ultimately, extinction. Extinction is just the end of a series of processes that began long before extinction occurred. The critical threshold was crossed when the remaining species populations were forced into the sequence of demographic processes leading to global extinction. As Goodman (1987) remarked, the loss of the last member of a species is always the result of bad luck. So it was for the second to the last member, and so on. A fundamental problem, then, is to find variables that mark the threshold along the sequence of demographic processes where decline becomes irreversible.

Counting species extinctions and near-extinctions is an ineffective approach to this problem. Politically, these counts are strong medicine, and they are fodder for vital discussions in moral philosophy. Scientifically, once we are convinced that the rate of species extinctions is unusually high and that our activities are causal, the numbers can be useful under only one circumstance. We would have to be able to trace back through the sequence of processes that compose an extinction and, by doing so, learn how to prevent the process from producing future extinctions. This can be successful only if we learn how to manipulate either the demographic processes leading to extinction or their causes. Important causes are often related to human activities and include economics, social values, politics, legislation, chance, or, most likely, some combination of these. Ultimately, except in cases of direct killing, the terminal process will be stochastic, evolutionary, or environmental. Except for nonadaptive genetic change enforced by us, these terminal causal agents will be either chance or environmental change.

The focus for biologists, then, is on the ecology of populations. Because the probabilities along the sequence of processes that can lead to biodiversity change are subject to political, social, economic, and legislative forces, these forces must be considered potent driving variables for the demographic processes of biodiversity change. Policy-makers and managers of biodiversity influence these drivers. Ecologists are responsible for revealing, to decision-makers and the public, the qualitative and quantitative nature of the interactions of these forces with population processes. The science of ecology cannot advise that some particular forest ecosystem is "best" except by reference to the context of society's prevailing system of values (Kimmins 1993). Policy often lacks real scientific content. The sole source of scientific content for improved policy is knowledge of environmental and demographic processes as well as knowledge of how human activities influence these processes. The resultant rates of species extinctions are no help here; this point is especially important for north temperate and northern environments.

My emphasis in this chapter is on vertebrate species, in which behavioural choices influence patterns of movement among spatially divided populations. Some of the principles apply to vascular plants, lichens, or fungi. Many of these, however, are dispersed by less predictable weather events such as windstorms, and their movements involve larger elements of chance.

The Nature of Biodiversity Changes

Changes in biodiversity at our latitudes do not result primarily from loss of species; they result from population degradation. The important point for policy and management is that species can suffer high rates of *population* (or local) extinctions before becoming threatened globally, and temperate species appear to suffer much higher rates of local population extinction before becoming endangered than do tropical species (e.g., Ehrlich and Daily 1993). Experimental population extinction is instructive in terms of our relationship to what we consider normal evolutionary process – the process we have relied on to save species from our actions (Henderson et al. 1985; Merriam 1991).

We commonly assume that if recruitment to a species' population lags behind its losses, selection pressures will act to correct that imbalance. We are taught that this change must take place at the level of the individual, otherwise we embrace the illogic of group selection. Broadly conceived, population regulation thus depends on excessive losses from the population causing effects down through social and spatial groups to the individual, where increased reproductive effect will correct the problem signalled by the selective force. It is clear, however, that by changing landscape structure we can prevent vital movements between subpopulations and cause the aggregated subpopulations that form the functional demographic unit to fail. With rapid landscape change, selective forces cannot save the demographic unit by changing individual reproductive output because response is too slow, but more rapid modification of behaviour and scale of movement can save the demographic unit (Merriam 1991). *Rapid manmade changes to the landscape, such as fragmentation, effectively move the response to selection pressure from the reproductive potential of individuals to the movement behaviour of individuals.* Selective forces still act on individuals, but they do so in response to powerful constraints from the larger scales above the individual. The effects of the spatial structure of the environment on demography are modulated by behaviour (e.g., Szacki et al. 1993; Bennett et al. 1994). Other behavioural adaptations, such as flexibility of habitat choice and ability to use anthropogenic novelties such as crops to supplement resources, can also provide an effective response to anthropogenic change (e.g., Wegner and Merriam 1990). Human technology can apply huge amounts of energy that change landscape structure, at rates never experienced in evolutionary history. The major question then is: how intense and how rapid can environmental changes be before they nullify evolutionary adaptations?

Landscapes in Heterogeneous Forest
Clearly our interest in biodiversity is caused by concern over the loss of valued elements of nature. Whether these elements are genetic variants, species, communities, or ecological landscapes, reducing the probability of their loss requires consideration of other parts of nature outside the individual elements. A primary question is how far outside the deme, or the population, or the floral or faunal assemblage, must we make measurements to be able to predict the fate of the subject of interest? We have recognized slowly and reluctantly, because of the complexity it implies, that the ecological future of any quadrat cannot be predicted from measurements taken only on that plot. Some "ecological neighbourhood" is required (Addicott et al. 1987; Pulliam et al. 1992). This is not simply a question of scale but a fundamental question of spatial and temporal distributions of ecological processes. It is generally accepted that changes to small regions or landscapes by humans can cause local elimination of a variant, species, or community type. Local elimination of a species starts with elimination of populations and subpopulations.

The requirement for spatial and temporal considerations of population processes arises from the condition of heterogeneity (e.g., Lord and Norton 1990; Fahrig and Merriam 1994). The mosaic created by forest harvesting has been generally recognized as a force of environmental change, but often has been considered a special case caused by fragmentation. This manmade patterning has been distinguished from natural processes except for comparisons of clearcut mosaics with those caused by forest fires and insect outbreaks (e.g., Hunter 1993). There is evidence (Middleton and Merriam 1983; Freemark and Merriam 1986; Krohne and Burgin 1990) that a heterogeneous mosaic is common and may be a normal property of continuous forests. Heterogeneity at several scales is identifiable in coastal, montane, and interior forests under the influences of, but not always congruent with, rock, soil, drainage, mesoclimate, and the overlay of secondary heterogeneity caused by biotic forces such as nitrogen injection by fixers, organic storage depots, or burrowers (Dwyer and Merriam 1981; Wiens 1989; Kolasa and Pickett 1991; Johnston 1995).

The lingering argument that some forests are less affected by heterogeneity because it occurs at a fine scale is becoming difficult to sustain even in the boreal forest. Harvesting history is imposing a specific mosaic structure over most of the boreal forest. Increasing knowledge shows that fires and insect infestations previously modified the boreal forest at different scales and in different ways but were probably as effective as harvesting in creating a variety of habitats but in variable patterns. The once-common notion that northern-latitude ecological systems were more homogeneous than southern-latitude ecological systems failed to consider the broader spatial scale in northern systems. The low and high Arctic provide clear examples of very large-scale linkage of resource patches into an annual resource array that sustains survival – a 4,400 km circuit for the Porcupine caribou herd and a link from the maritime littoral to the Arctic archipelago for the Inuit people. Until we have more evidence, the cautious position is that all forest systems are functional mosaics at some scale that is important to populations.

Environmental heterogeneity in the form of a mosaic, at a scale that matches resource uses, social structure, and movement behaviour of individuals, results in demographics that are understandable in terms of spatially divided population models. Several general constructs have been proposed recently for basic demographics, for population genetics, and for population management. Spatially divided population models include multipartite populations composed of interaction groups (Den Boer 1977, 1979), metapopulations (Levins 1970; Gilpin and Hanski 1991), island-mainland (Boorman and Levitt 1973), winking patches (Wilson 1980; Verboom et al. 1991), and source-sink, size-classed patches (Pulliam 1988; Hastings and Wollin 1989; Hastings 1991). Some scientists classify the models by their theoretical relationships rather than by the different ecological problems they address (e.g., Gotelli and Kelley 1993).

Animal populations commonly have functionally divisible populations in spatially separate habitat patches not only after fragmentation or patch-cutting but also in continuous forest (Krohne et al. 1984; Tomialojc et al. 1984; Wilcove 1988; Krohne and Burgin 1990). Variation in abundance in these patches, including local or patch extinction, causes abundance and recruitment gradients that will result in interpatch movements (Merriam and Wegner 1992; Villard et al. 1992). Juvenile dispersal, home range relocation, and disturbances add to the normal temporal array of movement across the landscape. Because animals must visit a temporally dynamic area of patches other than breeding patches to meet all their needs, classic static or territorial models are poor representatives of their use of space (Merriam 1990, 1991; Wegner and Merriam 1990; Bennett et al. 1994).

Taylor et al. (1993) discuss relationships among three measures of landscape structure and the four principal landscape processes. For any particular model of population dynamics, the *landscape structure* is a critical constraint (Dunning et al. 1992; Taylor et al. 1993; Fahrig and Merriam 1994). Critical *landscape processes* associated with these constraints can be used to learn about population changes that precede regional loss of species. A framework of four processes consists of: (1) landscape complementation, (2) landscape supplementation, (3) source-sink flows, and (4) neighbourhood flows. As proposed by Dunning et al. (1992), "complementation" refers to individuals getting their lifetime resource needs by visiting patches of different habitats. "Supplementation" refers to individuals using more than one patch of the same habitat type. "Source-sink flows" are movements of individuals among patches or spaces that differ in their ratio of immigration to emigration because of differences in net population production (Pulliam 1988). "Neighbourhood flows" are all the movements of other individuals and energy and matter in the functional ecological neighbourhood around individuals (cf. Addicott et al. 1987).

Variables of landscape structure constrain these processes. "Configuration" (or "physiognomy" of Dunning et al. [1992], or "pattern") measures the physical and geometric form of the elements of the landscape. Configuration can be subdivided into shape of patches (and inversely of matrix); size distribution of patches; size scale of patches; intrapatch stratification by size and shape; patterning by position effects such as runs of contiguity; patterning of patches by drainage, altitude, rock, and soil types; temporal dynamics of patch type in space (as in forest regrowth); third-dimensional patterning (as in forest canopy); patterning due to natural or built barriers; and more. Whereas configuration measures distributional qualities, the second variable, "composition," measures the qualitative elements that compose the pattern.

The third variable describing landscape structure arises from the first two. It is "connectivity" and is a measure of the probability of individuals moving through the landscape (Merriam 1984). It has two components, frequency of movement and frequency distribution of spatial scale of those movements, and is determined by species behaviour. Methods of measuring these parameters have been provided by Pielou (1975), Fahrig and Paloheimo (1988), O'Neill et al. (1988), Turner (1989), Henein and Merriam (1990), Gustafson and Parker (1992), Knaapen et al. (1992), Li and Reynolds (1993), and others. We have little knowledge of the meaning of particular values for any of these measurements. Until normal and abnormal values for particular environment types or management areas have been accumulated, our ability to measure exceeds our ability to interpret.

Population Processes Modified by Forest Practices

Forest practices modify the existing and resulting forest mosaics. Commonly forest practices are seen as "creating" the mosaic; probably they actually impose other mosaics over existing ones, which often are complex overlays from earlier disturbances. At the landscape scale, our knowledge of the historical dynamics of mosaic structure in forests is limited; Heinselman's (1973) heroic study dealt only with changes to the mosaic caused by fires. Such detailed work is unlikely to be repeated and, instead, will be replaced by remotely sensed data manipulated by geographic information systems with less emphasis on field observation (e.g., Pastor and Broschart 1990; Mladenoff et al. 1993).

This scale of data (and higher aggregates of it) is currently too coarse to explain changes caused to demographic processes in populations of many species (the root dynamics of biodiversity). This disparity will continue until we learn how to interpret from the larger scale to the more detailed, and from surrogates (such as vegetation) to processes (such as population dynamics). Remotely sensed data at coarse scales, however, are likely to be the most common basis for developing principles of predictable population changes when forestry alters environmental mosaics.

For species that are able to relocate to few (say, 20) breeding territories (or analogs) in the new average patch size, there is a high probability that their demographics will become characteristic of spatially divided populations (see also the appendix to this chapter). Under any of the theoretical constructs, spatially divided populations depend even more on an exchange of individuals among habitat patches. Models treating spatially divided populations are often lumped under the name *metapopulation,* although that is only one group of several classes of models for spatially divided populations (see above).

Levins's (1970) metapopulation model was among the earliest to deal with divided populations, and it is still a strong influence for some research groups (Gilpin and Hanski 1991; Gotelli and Kelley 1993). In the strict sense, metapopulation is a model derived from Levins's model. It assumes a multipatch population that depends for persistence on movement of individuals among patches that are of precisely equal environmental quality, without spatially explicit locations, each of which is equally accessible by all the individual organisms. Although several of the assumptions in the strict metapopulation model have been shown not to hold, the issue is how frequently and completely they are violated in forests. At the coarse level (large areas and broadly defined patch types), statistical predictions of such models may hold (e.g., Lankester et al. 1991; Verboom et al. 1991). In a managed forest of many more finely scaled patches, the predictions are less likely to hold. Either the characteristics of the mosaic or the number of patches may require a spatially explicit approach that recognizes the environmental inequality of patches – their unequal accessibility to individuals both seasonally and because of the temporal dynamics of patch quality in space. The following discussion invokes an unspecialized approach to spatially divided populations with a management objective of safeguarding a variety of populations following mosaic modification.

The spatial mosaic will cause heterogeneous distributions of populations at different spatial scales (Andrewartha and Birch 1984; Wiens 1989; and others). For our discussion we need two of those scales. "Local population" means the smallest population inhabiting a discrete habitat patch, often called a "patch population." "Regional population," often called a "landscape population," is an aggregate of patch populations from an arbitrary small region such as a small watershed or a management unit. It may be possible to map a real boundary for a patch population, but only by using a surrogate for the discrete habitat patch (usually the vegetation).

A critical feature in spatially divided populations is that all the local or patch populations interact by movement of individuals within a larger demographic network, or broader scale of the regional or landscape population. This latter population has no boundary unless one is enforced by a major barrier; the population is entirely conceptual (Merriam et al. 1989; Merriam 1991). The concept is fundamental to all theory of spatially divided populations; if we study individual patch populations, we find that they come and go by local extinctions and recolonizations in a stochastic manner. There is no predictable pattern to the fate of individual small-patch populations. That is why one model is called "winking patches" – they wink on and off like random lights. To obtain predictability about population survival, therefore, we increase the scale by including more patches in the sample until the sample stops behaving stochastically. The regional population is persistent even though the patch populations are ephemeral (Fahrig and Merriam 1994). Metapopulations or any other spatially divided demographic units do not have boundaries. They are conceptual units and, for measurements in the field, like any other sample, they exist to represent something larger and to let us deal with generalities that we cannot conveniently measure.

If we examine the environmental mosaic at a scale larger than the sample needed to represent the regional population, we can map population processes even though there are no firm boundaries. By knowing frequencies of movements between pairs of patch populations, or estimating their probabilities, we can estimate lines of isoflux (cf. Merriam 1991). Flux among patches will go to zero at absolute barriers and will form a gradient of increasing flux across lesser barriers such as environments created in the mosaic by forest management or other human activities. Fluxes of individuals are not necessarily proportional to distance because they are behaviourally determined. Flux probabilities will not necessarily be symmetrical between patches (creating the possibility of "valves"), seasonally constant, or homogeneous among sex, age, or behavioural groups (creating the possibility of filters for population structure) (Merriam et al. 1989).

As behavioural and environmental isolation of patch populations increase, flux or movements will decrease, regional populations may become isolated, and the probabilities of both local and regional extinctions will increase. This is the beginning of larger, potentially global extinction unless the accumulation of regional extinctions is halted at the finer scale. If, instead of an advancing wave of regional extinctions, the pattern is of expanding gaps of regional extinctions within the biogeographic range of the species, predicting the global result may be delayed long enough so that restoration is neither practical nor possible.

If, in contrast, members of the population have, or acquire by adaptation, movement abilities that let them move freely across non-habitat patches in the mosaic, patchiness can be overcome by the movement behaviour. The capacity for movement makes the scale of the mosaic finer and allows individuals to treat the heterogeneity as though it were not patchy. At the opposite extreme, where a patch contains enough of all the vital resources to support a large population and encompasses the extreme movement range of the organism, the population will see the environment as uniform and not patchy. This threshold between patchiness and uniform heterogeneity, assessed as evenness (Pielou 1975; Allen and Hoekstra 1991), is determined by movement and cannot be measured using biophysical surrogates, such as vegetation type, without a knowledge of movement behaviours and a knowledge of the abilities of particular species and even of classes of individuals in a species.

For the same reasons, the scale at which questions of population survival are examined must be validated by measurements of movement of individuals among the critical patches of the mosaic. The interaction between this scale of ecological neighbourhood and the scale at which our activities modify the landscape is a potent area for effectively managing biodiversity. Because predictions of both demography and genetics of spatially divided populations depend on the interconnection of subpopulations by movements of individuals, information about landscape movements must be part of the primary database for management planning. Conversely, modifications of landscape composition and/or configuration, both of which can affect connectivity, must be assumed to have direct effects on the survival of populations until otherwise demonstrated.

The fashionable recommendation of corridors as management interventions often focuses entirely on the movement component of demographic processes in spatially divided populations. This and similar recommendations commonly assume that subjectively sited structural interconnections of habitat patches with corridors of similar vegetation will restore normal demographic and genetic processes to a mosaic. Demographic processes in a mosaic are changed by:

- elimination of habitat (sometimes most of it)
- change of landscape composition (removal of all or disproportionate amounts of some resource types)
- reconfiguration of the remaining resource patches
- qualitative and quantitative addition of disturbances
- selective adjustment of species proportions in communities (including addition of exotics and crops) with consequent readjustment of species interactions
- modification of biophysical environmental variables such as mesoclimate and pollutants.

Advocating movement corridors alone to restore normalcy to such a potential array of factors is simplistic. Simulations suggest that population persistence is encouraged more by increased connectivity (as opposed to arbitrarily inserting corridors) than by adjusting demographic rates related to reproductive potential (Fahrig and Paloheimo 1988). Corridors that do increase connectivity should slow the process of isolation of subpopulations and subsequent accumulation of local extinctions.

Restoring primeval population processes, however, would mean first restoring some form of original mosaic in which connectivity was incorporated in the composition and configuration of the complete mosaic. In this scenario corridors are not necessary; any interconnected set of patches will serve if the animals will choose them for movement. Restoring population processes to some threshold of predictable population persistence will require structuring the landscape to some point on the gradient between this primeval condition and the severely fragmented landscape for which corridors are often suggested as a mitigating measure.

Forest management inevitably changes some of the six variables affecting population processes listed above. Because it changes heterogeneity at some scale, it should always be assumed to modify landscape structure for some species. We should be able to control how much these changes affect the probabilities of population persistence by adjusting the degree of change of connectivity when the landscape is being restructured. Rather than structuring corridors into the post-harvest mosaic, it would be more effective to plan the harvest and silviculture to leave natural spatial units with high ecological productivity and resource values that also interconnect regrowth patches (Figure 4.1).[1]

---

1   As illustrated in Figure 4.1, connectivity is encouraged in a fashion very similar to the Forest Ecosystem Network of the British Columbia Forest Practices Code.

Connectivity results from landscape composition and landscape configuration. Examples of natural ecological units with high connectivity value, at several scales, include montane valley bottoms, riparian zones (Bunnell

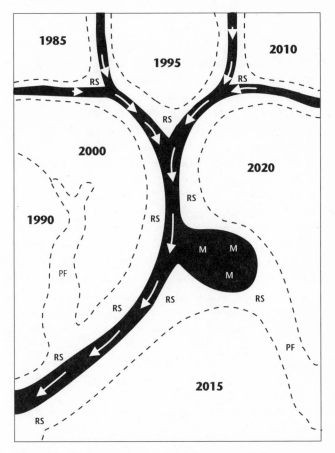

Figure 4.1 An approach to maintaining connectivity in a managed forest by exploiting natural landscape features. Forest harvest areas are indicated by year of harvest. Previously harvested areas are expected to be restocking and recolonizing with species that moved because habitat was removed. Marsh (M), riparian forested strips (RS), and poorly drained forest (PF) are all known to provide some combination of high plant productivity, special habitats, landscape-scale environmental protection, and below-average forest economic yield. Such patches could remain unharvested both for their ecological values and to enhance interconnection and recolonization of regrowing harvested areas. These interconnections are natural ecological features and follow natural landscape patterns such as topographic lows and drainageways (– – –).

and Dupuis 1995), avalanche tracks, and poorly drained topographic networks that often head drainage systems and have high primary and secondary productivity (Sjöberg 1989). By using ecological resource patches and drainage patterns (as in Figure 4.1) instead of the geometric shapes and artificial boundaries of commerce and engineering, landscape connectivity could be improved at the fundamental level of landscape composition and configuration.

In historically mismanaged mosaics, restoring connectivity may require less fundamental but more practicable adjustments, such as inserting movement and habitat corridors (Harris and Gallagher 1989; Bennett 1991; Harrison 1992) to buy time until more fundamental connectivity can be restored. Criticisms of corridors (e.g., Simberloff and Cox 1987; cf. Noss 1987; Hobbs 1992; Simberloff et al. 1992) speculate on the dangers of increased invasion by pathogens and increased predation due to prey concentration along corridors, and on the costs of purchasing extensive interconnecting land areas. The direct empirical basis for pathogen invasion and predator traps as it might apply to forested landscapes is possibly weaker than the direct evidence that animals do follow corridors (e.g., Merriam and Lanoue 1990). We know from simulations that moderately increased mortality in poor-quality corridors can significantly affect the whole spatially divided population (Henein and Merriam 1990). The assumption that mortality will be proportional to distance travelled in corridors is doubtful. Merriam and Lanoue (1990) showed that some mammals spent 84% of their corridor transit time under cover and used only 16% of their time in bursts of movement. Knowledge of animal movements in corridors is far from complete, but it is sufficient to allow constructive criticism with results from empirical or model tests. We have little theoretical basis for speculation, and empirical studies are required.

Lack of definitive proof that organisms follow corridors and that corridors save populations has also been cited as evidence against recommending corridors. Data that directly show corridor use are most common, of course, for easily studied species such as small mammals and birds (e.g., Wegner and Merriam 1979; Merriam and Lanoue 1990; Bennett et al. 1994). Inference of corridor use is more common (e.g., Henderson et al. 1985; Szacki 1987; Harris and Gallagher 1989; Bennett 1990, 1991; Villard et al. 1992). Suggestions that corridors are effective for plant movements are unconvincing (Fritz and Merriam 1993; Matlack 1994; but cf. Burel and Baudry 1990; Noss 1993). Demonstrations that corridors can affect population persistence come from simulations (Fahrig and Merriam 1985; Henein and Merriam 1990). Empirical demonstration that corridors save populations awaits results from experimental model systems (Wiens et al. 1993; Ims et al. 1993). Demonstrations that may be convincing but that will not show causality may also arise from management attempts, based mainly on simulation models, such as those for Netherlands badgers (Lankester et al. 1991).

The issue of costs emerges from socio-legal structures; it is really an issue of values, not a scientific one. Scientific advice and results of management tests, however, will form important components of decision-making tools and policy judgments. This volume is part of that process.

Management of Forest Biodiversity

Managers should expect, at best, a comprehensive checklist of questions that they must answer for their area and their problem before proceeding with any intervention. The appendix to this chapter provides one example. Often, getting the answers will require an experimental application of the proposed intervention with half of the area held as a control. The managers will need to apply adaptive management as part of the normal management program (Holling et al. 1986; Merriam et al. 1992).

The special knowledge managers possess of their ecological area, target species, and cultural history makes the role of the local manager especially valuable in any management program and its research base. Local managers, using local cultural knowledge, should attempt "backcasting" the structure and processes of the environmental mosaic to provide a quasi-baseline with which to compare the current ecological state and trends (e.g., Vernier 1995).

In applying this knowledge on the land, we must emphasize that simplistic, all-purpose answers are not what will be supplied to managers. Ecological systems are characterized by variance and by complexities of alternative states that make them unlike engineered systems and make demands for "handbook answers" dangerously naive. The interaction of social values and scientific knowledge in management plans also needs realistic consideration. Indicators, target species, and landscape features that are incorporated into management applications are often heavily influenced by values that lie outside the effective realm of science.

Ecological Principles Recommended for Biodiversity Policy Development

Uncertainty will always characterize the current best scientific assessment. We must make some policy decisions about conserving biological resources before that uncertainty is reduced to the scientific norm. The scientific norm safeguards against finding a relationship where none exists. In many conservation policy issues, we must reduce that safeguard and take more risk of finding a false cause, because if the effect is real and we wait until evidence reduces the risk of false positives to the norm, we will lose many natural features. This "precautionary principle" requires us to offer ecological

principles rather than rules to guide policy development. The following principles are offered in that context; they are also a summary of this
chapter.

(1) Changes in forest biodiversity result ultimately through demographic
    processes.
(2) Demographic processes in forest species have important spatial and
    temporal dimensions.
(3) All forestry practices change the complex of spatial and temporal di-
    mensions of the mosaics of forest environments.
(4) A common effect of forest practices is to spatially divide populations.
(5) Spatially divided populations require an environmental mosaic that
    will supply their complete lifetime array of vital resources, including
    mates, food, and shelter.
(6) Spatially divided populations often depend for survival on movements
    of individuals between subpopulations so that local (patch) extinctions
    do not accumulate.
(7) In a mosaic of various stages of succession, spatially divided populations
    may depend, for continued access to required patches, on movements
    of individuals from source patches to colonization sites.
(8) The potential effects of forest practices on forest biodiversity can best
    be predicted or assessed in a framework of landscape structure that
    consists of composition, configuration, and connectivity.
(9) Effects on forest biodiversity can be reduced if the effects of forest prac-
    tices on landscape structure emulate the natural environmental mo-
    saic in the managed mosaic.
(10) Answers will not be supplied by research ecologists in detail suitable
    for site-specific management problems. Consequently, most manage-
    ment programs will need to include an experimental research compo-
    nent.
(11) Because complete answers will not be available for most problems, adap-
    tive management, monitoring, and adjustment will be required.

Appendix:
Some Questions Underlying the Persistence of Species in Forests
These questions illustrate the relationships governing the persistence of
species in forests. The list is longer than a practicable checklist sought by
managers, but it serves to summarize the kinds of relationships that are
important. The questions are grouped according to the components of landscape structure discussed in this chapter: composition, configuration, and
connectivity.

Landscape Composition

(1) Have humans appropriated so much of local net primary production that the life support system for other heterotrophs will fail (Vitousek et al. 1986)?

(2) Do so many resource patches have toxic qualities (pollutants) that the landscape may be a deathtrap?

(3) Has the environmental quality of many patches been reduced to such a condition that populations cannot achieve net recruitment, making them sinks that draw off the recruitment from source populations (Pulliam 1988)?

(4) Are all resource needs, including mates and ephemeral needs, represented in the landscape composition?

(5) If the management area is recovering through natural recolonization, is the scale of planning large enough to accommodate multiple-source populations and movements from them? (See also "Connectivity.")

(6) Are patch populations spatially fixed or do they move among resource patches? If they move among resource patches, what is the rate of flow of populations through important patch types in each season (Wegner and Merriam 1990)?

(7) Do populations flow through a large number of resource patch types, or are they constrained to a few specific resource types? (Note that some population data, especially those for birds, represent only breeding habitat and often do not indicate other resource needs or uses.)

(8) If populations are flowing to one or a few resource types, are these distant from the rest of the array of resource patches? (If so, consider connectivity requirements, below, or shift the scale of the management unit up until it takes in the distant resource patch[es].)

(9) Is the productivity of source patches sufficient to compensate for stochastic processes in recipient patches? Do sinks or populations in submarginal habitat serve as backup to recolonize source populations after stochastic failures? Does this process outweigh the mortality in the sinks between stochastic failures of source populations?

(10) Have exotic patch types been added to the mosaic, and do they act as novel resource centres, altering original ecological processes?

Landscape Configuration

(1) Has landscape configuration caused spatial division of populations?

(2) Is the configuration of resource patches dictated by natural boundaries (such as dendritic drainage or altitude stratification) or by anthropogenic patterns (such as cut blocks [Franklin and Forman 1987], roads, or refuge boundaries [Paquet 1991])? (See also "Connectivity" below.)

(3) Is the array of required resource patches separated by environments that can render needed resources unavailable or variable in their availability? (See also [8] and [10] under "Landscape Development.") If so, does the distance to the needed resources exceed the movement range of the species (Fahrig and Paloheimo 1988)? If distance is not the barrier but probability of access is lowered by behavioural reluctance to make the move, does attempted movement increase mortality and affect population survival (Horejsi 1989; Henein and Merriam 1990; Paquet 1991)? If mortality is not seriously increased but behavioural reluctance lowers the availability of a needed resource, does this significantly decrease landscape quality for the population (Merriam et al. 1989)?

(4) Are population patches subject to alien influences such as exotic animals (as in Wood Buffalo National Park), exotic herbs (as in Pukaskwa National Park), hunters using logging roads, or noise?

(5) Is the area of resource 20 times^2 the expected area of a breeding territory for the target species? (If so, and if environmental changes caused by penetration of the patch edge by outside conditions reduce habitat selection in the edge, this could reduce breeding territories to 10 or fewer. This condition may invoke stochastic events as the main forces governing patch population dynamics as well as significantly reducing the landscape population [Ambuel and Temple 1983; Saunders et al. 1991]. Note that data for reduced environmental quality in edges, including increased edge predation, come mainly from forest interior songbirds in North America [Ambuel and Temple 1983, but see also Yahner 1988; Laurance and Yensen 1991] and from corvid predators in Scandinavia. It is conclusive only for a few northern forest situations [Andren and Angelstam 1988].)

(6) Are areas of all patches of breeding habitat small, as in (5) above? (If so, and if the modal interpatch distance is much less than the average length of a movement by an individual, some species may adapt their habitat selection behaviour and use smaller patches [Middleton and Merriam 1983; Freemark 1988]. Effects on extinction probabilities are unclear.)

(7) Are the edge effects noted in (5) above likely? (They are increasingly likely as the perimeter:area ratio exceeds four times the minimum dimension of a patch.)

(8) If perimeter:area ratios are large, regardless of patch area, do the shapes of edges interact with movement behaviour to influence rates and locations of interpatch movements? (See "Connectivity" below.)

---

2 The figure 20 is necessarily an arbitrary, though reasoned, estimate. Actual vulnerability to potential edge effects is undoubtedly species-specific.

Connectivity

(1) Is the larger, landscape population of the target species spatially divided? If so, what is the modal size of the seasonal minimum population per patch? (If it is less than about 20, the dynamics of the patch populations are likely to be driven by stochastic events and therefore to be unpredictable on a patch-by-patch basis [Fahrig and Merriam 1994].)

(2) Is the modal movement distance of a species about equal to interpatch distances between vital resources? (If so, and if spatially divided populations are small, as in [1] above, improving connectivity may increase productivity and survival of the larger, landscape population. Note that normal connectivity occurs through an unmodified mosaic of resource patches; connectivity that is dominated by anthropogenic features, such as hedges or fencerows, is characteristic of landscapes with dysfunctional mosaics [Merriam and Saunders 1993]. Note also that connectivity can depend on any type of resource patch, including unvegetated spaces and built structures. The movement behaviour of the species dictates the mix, and this may vary for a species according to behavioural adaptation in particular landscapes [Merriam and Lanoue 1990; Bennett et al. 1994].)

(3) Does the perimeter shape of spatially divided populations control connectivity spatially (Laurance and Yensen 1991)?

(4) Is connectivity for the landscape population limited by connectivity for one particular sex, age, or life stage in a particular season?

(5) Is connectivity predictable only from movement data, or can it be predicted from map features and vegetation types (Villard et al. 1992; Bennett et al. 1994)?

(6) Is the low connectivity for many seed plants likely to become a long-term limitation to connectivity for animals in the landscape, or are animals dispersing plants enough to avoid this limitation (Fritz and Merriam 1993; Young et al. 1993; Matlack 1994)?

Literature Cited

Addicott, J.F., J.M. Aho, D.F. Antolin, J.S. Richardson, and D.A. Soluk. 1987. Ecological neighbourhoods: Scaling environmental patterns. Oikos 49:340-46.

Allen, T.F.H., and T.W. Hoekstra. 1991. Role of heterogeneity in scaling of ecological systems under analysis. Pp. 45-68 *in* J. Kolasa and S.T.A. Pickett (eds.). Ecological heterogeneity. Springer-Verlag Ecological Studies 86, New York, NY.

Ambuel, B., and S.A. Temple. 1983. Area-dependent changes in the bird communities and vegetation of southern Wisconsin forests. Ecology 64:1057-68.

Andren, H., and P. Angelstam. 1988. Elevated predation rates as an edge effect in habitat islands. Ecology 69:544-47.

Andrewartha, H.G., and L.C. Birch. 1984. The ecological web. University of Chicago Press, Chicago, IL.

Bennett, A.F. 1990. Habitat corridors and the conservation of small mammals in a fragmented forest environment. Landscape Ecology 4:109-22.

–. 1991. Habitat corridors, their role in wildlife management and conservation. Victoria Department of Conservation and Environment, Arthur Rylah Institute for Environmental Research, Heidelberg, Australia.

Bennett, A.F., K. Henein, and G. Merriam. 1994. Corridor use and the elements of corridor quality: Chipmunks and fencerows in an agricultural mosaic. Biological Conservation 68:155-65.

Boorman, S.A., and P.R. Levitt. 1973. Group selection on the boundary of a stable population. Theoretical Population Biology 4:85-128.

Bunnell, F.L., and L. Dupuis. 1995. Riparian habitats in British Columbia: Their nature and role. Pp. 7-21 *in* K.H. Morgan and M.A. Lashmar (eds.). Riparian habitat management and research. A special publication of the Fraser River Action Plan, Canadian Wildlife Service, Delta, BC.

Burel, F., and J. Baudry. 1990. Hedgerow networks as habitats for forest species: Implications for colonizing abandoned agricultural land. Pp. 82-97 *in* R.G.H. Bunce and D.C. Howard (eds.). Species dispersal in agricultural habitats. ITE and Belhaven Press, London, UK.

Den Boer, P.J. 1977. Dispersal power and survival: Carabids in a cultivated countryside. Landbouwhoggeschul Wageningen Miscellaneous Papers 14:1-190.

–. 1979. The significance of dispersal power for the survival of species, with special reference to the carabid beetles in a cultivated countryside. Fortschrift für Zoologie 25:79-94.

Dunning, J.B., J.B. Danielson, and H.R. Pulliam. 1992. Ecological processes that affect populations in complex landscapes. Oikos 65:169-75.

Dwyer, L.M., and G. Merriam. 1981. Influence of topographic heterogeneity on deciduous leaf litter decomposition. Oikos 37:228-37.

Ehrlich, P.R., and G.C. Daily. 1993. Population extinction and saving biodiversity. Ambio 22:64-68.

Fahrig, L., and G. Merriam. 1985. Habitat patch connectivity and population survival. Ecology 66:1762-68.

–. 1994. Conservation of fragmented populations. Conservation Biology 8:50-59.

Fahrig, L., and J. Paloheimo. 1988. Determinants of local population size in patchy habitats. Theoretical Population Biology 34:194-213.

Franklin, J.F., and R.T.T. Forman. 1987. Creating landscape patterns by forest cutting: Ecological consequences and principles. Landscape Ecology 1:5-18.

Freemark, K.E. 1988. Agricultural disturbance, wildlife and landscape management. Pp. 77-84 *in* M. Moss (ed.). Landscape ecology and management. Polyscience, Montreal, PQ.

Freemark, K.E., and G. Merriam. 1986. Importance of area and habitat heterogeneity to bird assemblages in temperate forest fragments. Biological Conservation 36:115-41.

Fritz, R., and G. Merriam. 1993. Fencerow habitat for plants moving between farmland woods: An assessment of corridor potential. Biological Conservation 64:141-48.

Gilpin, M., and I. Hanski. 1991. Metapopulation dynamics: Brief history and conceptual domain. Biological Journal of the Linnean Society 42:1-16.

Goodman, D. 1987. The demography of chance extinction. Pp. 11-34 *in* M. Soulé (ed.). Viable populations for conservation. Cambridge University Press, Cambridge, UK.

Gotelli, N.J., and W.G. Kelley. 1993. A general model of metapopulation dynamics. Oikos 68:36-44.

Gustafson, E.J., and G.R. Parker. 1992. Relationships between landcover proportion and indices of landscape spatial pattern. Landscape Ecology 7:101-10.

Harris, L.D., and P.B. Gallagher. 1989. New initiatives for wildlife conservation: The need for movement corridors. Pp. 11-34 *in* Preserving communities and corridors. Defenders of Wildlife, Washington, DC.

Harrison, R.L. 1992. Towards a theory of inter-refuge corridor design. Conservation Biology 6:293-95.

Hastings, A. 1991. Structured models of metapopulation dynamics. Biological Journal of the Linnean Society 42:57-71.

Hastings, A., and C. Wolin. 1989. Within-patch dynamics in a metapopulation. Ecology 70:52-71.

Heinselman, M.L. 1973. Fire in the virgin forests of the Boundary Waters Canoe Area, Minnesota. Quaternary Research 3:329-82.

Henderson, M.T., G. Merriam, and J. Wegner. 1985. Patchy environments and species survival: Chipmunks in an agricultural mosaic. Biological Conservation 31:95-105.

Henein, K., and G. Merriam. 1990. The elements of connectivity where corridor quality is variable. Landscape Ecology 4:157-70.

Hobbs, R.J. 1992. The role of corridors in conservation: Solution or bandwagon. Trends in Ecology and Evolution 7:389-92.

Holling, C.S., G.B. Dantzig, and C. Winkler. 1986. Determining optimal policies for ecosystems. Pp. 453-73 *in* M. Kallio, A.E. Anderson, R. Seppala, and A. Morgan (eds.). Systems analysis in forestry and forest industries. Elsevier, Amsterdam, Netherlands.

Horejsi, B.L. 1989. Uncontrolled land-use threatens an international grizzly bear population. Conservation Biology 3:220-23.

Hunter, M.L. 1993. Natural fire regimes as spatial models for managing boreal forests. Biological Conservation 65:115-20.

Ims, R.A., J. Rolstad, and P. Wegge. 1993. Predicting space use responses to habitat fragmentation: Can voles *Microtus eoconomus* serve as an experimental model system (EMS) for capercaillie grouse in boreal forest? Biological Conservation 63:261-68.

Johnston, C. 1995. Effects of animals on landscape pattern. Pp. 57-80 *in* L. Hansson, L. Fahrig, and G. Merriam (eds.). Mosaic landscapes and ecological processes. Chapman and Hall, London, UK.

Kimmins, J.P. 1993. Ecology, environmentalism and green religion. Forestry Chronicle 69:285-89.

Knaapen, J.P., M. Scheffer, and B. Harms. 1992. Estimating habitat isolation in landscape planning. Landscape and Urban Planning 23:1-16.

Kolasa, J., and S. Pickett (eds.). 1991. Ecological heterogeneity. Springer-Verlag, New York, NY.

Krohne, D.T., and A.B. Burgin. 1990. The scale of demographic heterogeneity in a population of *Peromyscus leucopus*. Oecologia 82:97-101.

Krohne, D.T., B.A. Dubbs, and R. Baccus. 1984. Analysis of dispersal in an unmanipulated population of *Peromyscus leucopus*. American Midland Naturalist 112:146-56.

Lankester, K., R.C. van Apeldoorn, E. Meelis, and J. Verboom. 1991. Management perspectives for populations of the Eurasian badger *Meles meles* in a fragmented landscape. Journal of Applied Ecology 28:561-73.

Laurance, W.F., and E. Yensen. 1991. Predicting the impacts of edge effects in fragmented habitats. Biological Conservation 55:77-92.

Levins, R. 1970. Extinction. Pp. 77-107 *in* M. Gerstenhaber (ed.). Lectures on mathematics in the life sciences, Vol. 2. American Mathematical Society, Providence, RI.

Li, H., and J.F. Reynolds. 1993. A new contagion index to quantify spatial patterns of landscapes. Landscape Ecology 8:155-62.

Lord, J.M., and D.A. Norton. 1990. Scale and the spatial concept of fragmentation. Conservation Biology 4:197-202.

Matlack, G. 1994. Plant species migration in a mixed-history forest landscape in eastern North America. Ecology 75:1491-1502.

Merriam, G. 1984. Connectivity: A fundamental ecological characteristic of landscape pattern. Pp. 1-15, Theme I *in* J. Brandt and P. Agger (eds.). Proceedings, First International Seminar on Methodology in Landscape Ecological Research and Planning. International Association for Landscape Ecology, Roskilde University, Roskilde, Denmark.

–. 1990. Ecological processes in the space and time of farmland mosaics. Pp. 121-33 *in* I.S. Zonneveld and R.T.T. Forman (eds.). Changing landscapes: An ecological perspective. Springer-Verlag, New York, NY.

–. 1991. Corridors and connectivity: Animal populations in heterogeneous environments. Pp. 134-42 *in* D. Saunders and R. Hobbs (eds.). The role of corridors in nature conservation. Surrey Beatty and Sons, Chipping Norton, NSW, Australia.

Merriam, G., and A. Lanoue. 1990. Corridor use by small mammals: Field measurements for three experimental types of *Peromyscus leucopus*. Landscape Ecology 4:123-31.

Merriam, G., and D. Saunders. 1993. Corridors in restoration of fragmented landscapes. Pp. 71-87 *in* D. Saunders, R. Hobbs, and P. Ehrlich (eds.). Nature Conservation 3: Reconstruction of fragmented ecosystems, global and regional perspectives. Surrey Beatty and Sons, Chipping Norton, NSW, Australia.

Merriam, G., and J. Wegner. 1992. Local extinctions, habitat fragmentation and ecotones. Pp. 150-69 *in* A.J. Hansen and F. Di Castri (eds.). Landscape boundaries. Ecological Studies, Springer-Verlag, New York, NY.

Merriam, G., M. Kozakiewicz, E. Tsuchiya, and K. Hawley. 1989. Barriers as boundaries for metapopulations and demes. Landscape Ecology 2:227-35.

Merriam, G., J. Wegner, and S. Pope. 1992. Parklands: Parks in their ecological landscapes. Report prepared by the Carleton University Landscape Ecology Laboratory for Canadian Parks Service. Ottawa-Carleton Institute of Biology, Carleton University, Ottawa, ON.

Middleton, J.D., and G. Merriam. 1983. Distribution of woodland species in farmland woods. Journal of Applied Ecology 20:625-44.

Mladenoff, D.J., M.A. White, J. Pastor, and T.R. Crow. 1993. Comparing spatial pattern in unaltered old-growth and disturbed forest landscapes. Ecological Applications 3:294-306.

Noss, R. 1987. Corridors in real landscapes: A reply to Simberloff and Cox. Conservation Biology 1:159-64.

–. 1993. Wildlife corridors. Pp. 43-68 *in* D.S. Smith and P.C. Hellmund (eds.). Ecology of greenways. University of Minnesota Press, Minneapolis, MN.

O'Neill, R.V., J.R. Krummel, R.H. Gardner, G. Sugihara, B. Jackson, D.L. DeAngelis, B.T. Milne, M.G. Turner, B. Zygmunt, S.W. Christensen, V.H. Dale, and R.L. Graham. 1988. Indices of landscape pattern. Landscape Ecology 1:153-62.

Paquet, P.C. 1991. Black bear ecology in the Riding Mountains. Final Report. Prepared for Manitoba Natural Resources and Canadian Park Service by John/Paul & Associates.

Pastor, J., and M. Broschart. 1990. The spatial pattern of a northern conifer-hardwood landscape. Landscape Ecology 4:55-68.

Pielou, E.C. 1975. Ecological diversity. John Wiley and Sons, New York, NY.

Pulliam, R.E. 1988. Sources, sinks and population regulation. American Naturalist 132:652-61.

Pulliam, H.R., J.B. Dunning, and J. Liu. 1992. Population dynamics in complex landscapes: A case study. Ecological Applications 2:165-77.

Sagoff, M. 1993. Biodiversity and the culture of ecology. Bulletin of the Ecological Society of America 74:374-81.

Saunders, D.A., R.J. Hobbs, and C.R. Margules. 1991. Biological consequences of ecosystem fragmentation: A review. Conservation Biology 5:18-32.

Simberloff, D., and J. Cox. 1987. Consequences and costs of conservation corridors. Conservation Biology 1:63-71.

Simberloff, D., J.A. Farr, J. Cox, and D.W. Mehlman. 1992. Movement corridors: Conservation bargains or poor investments? Conservation Biology 6:493-504.

Sjöberg, K. 1989. Habitat changes in boreal coniferous forest. Pp. 19-21 *in* The importance of residual biotopes for flora and fauna. Swedish Environmental Protection Board, Solna, Sweden.

Szacki, J. 1987. Ecological corridors as a factor in determining structure and organization of bank vole populations. Acta Theriologica 32:31-44.

Szacki, J., J. Babinska-Werka, and A. Liro. 1993. The influence of landscape spatial structure on small mammal movements. Acta Theriologica 38:113-23.

Taylor, P.D., L. Fahrig, K. Henein, and G. Merriam. 1993. Connectivity is a vital element of landscape structure. Oikos 68:571-73.

Tomialojc, L., T. Wesolowski, and W. Walankiewicz. 1984. Breeding bird community of a primeval temperate forest (Bialowieza National Park, Poland). Acta Ornithologica 20:251-311.

Turner, M.G. 1989. Landscape ecology: The effect of pattern on process. Annual Review of Ecology and Systematics 20:171-97.

Verboom, J., K. Lankester, and J. Metz. 1991. Linking local and regional dynamics in stochastic metapopulation models. Biological Journal of the Linnean Society 42:39-55.

Vernier, P. 1995. Effects of clearcutting on landscape structure and bird species diversity and abundance in the Rocky Mountains. M.Sc. thesis, University of British Columbia, Vancouver, BC.

Villard, M-A., K.E. Freemark, and G. Merriam. 1992. Metapopulation dynamics as a conceptual model for neotropical migrant birds: An empirical investigation. Pp. 474-82 *in* J.M. Hagan and D.W. Johnston (eds.). Ecology and conservation of neotropical migrant landbirds. Smithsonian Institution Press, Washington, DC.

Vitousek, P.M., P.R. Ehrlich, A.H. Ehrlich, and P.A. Matson. 1986. Human appropriation of the products of photosynthesis. BioScience 36:368-73.

Wegner, J., and G. Merriam. 1979. Movements by birds and small mammals between a wood and adjoining farmland. Journal of Applied Ecology 16:349-57.

–. 1990. Use of spatial elements in a farmland mosaic by a woodland rodent. Biological Conservation 54:263-76.

Whittaker, R.H. 1972. Evolution and the measurement of species diversity. Taxon 21:213-51.

Wiens, J.A. 1989. Spatial scaling in ecology. Functional Ecology 3:385-97.

Wiens, J.A., N.C. Stenseth, B. Van Horne, and R.A. Ims. 1993. Ecological mechanisms and landscape ecology. Oikos 66:369-80.

Wilcove, D.S. 1988. Changes in the avifauna of the Great Smoky Mountains: 1947-1983. Wilson Bulletin 100:256-71.

Wilson, D.S. 1980. The natural selection of populations and communities. Benjamin/ Cummings, Menlo Park, CA.

Wilson, E.O., and F.M. Peters. 1988. Biodiversity. National Academy, Washington, DC.

WRI/IUCN/UNEP. 1992. Global Biodiversity Strategy. World Resources Institute/International Union for Conservation of Nature and Natural Resources/United Nations Environment Program, Gland, Switzerland.

Yahner, R.H. 1988. Changes in wildlife communities near edges. Conservation Biology 2:333-39.

Young, A.G., S.I. Warwick, and G. Merriam. 1993. Genetic variation and structure at three spatial scales for *Acer saccharum* (sugar maple) in Canada and the implications for conservation. Canadian Journal of Forest Research 23:2568-78.

5
Measuring Diversity of Communities and Ecosystems with Special Reference to Forests
Daniel Simberloff

Introduction

Measuring diversity of communities and ecosystems requires first that we determine what communities and ecosystems are. Communities are collections of species, but we have as yet arrived at no consensus on the degree to which these collections are organized. Ecosystems are communities plus their physical environments, and controversy exists over the nature of ecosystems just as controversy exists over the nature of communities.

Numerous traits of communities and ecosystems can be construed as contributing to diversity. For some of these traits (such as species richness or turnover rates) we have well-established statistics believed to express "diversity"; for other traits there has been little attempt at quantification or there is little agreement on what measures are appropriate. The term *biodiversity*, however, has come to mean more than community or ecosystem diversity for any of these traits, or even for all of them together. *Biodiversity* transcends the community and ecosystem levels, even though these levels are probably the locus of most features that have traditionally been seen as contributing to biodiversity.

Because the nature of communities and ecosystems is not well understood and none of the traditional diversity statistics adequately captures the full scope of "biodiversity," what exactly do we measure and monitor? To some extent this question can be clarified by asking ourselves what we want from biodiversity and what aspects of biodiversity we wish to conserve. Then we must ask what we can monitor that will help us manage for those aspects. Clearly economics alone dictates that we cannot monitor every community and ecosystem trait that might be of interest. So we are forced to take shortcuts by monitoring components or aspects of the community and ecosystem whose status also tells us about other features of interest to us.

We can consider two broad possibilities. First, we may be able to establish real correlations between certain components and features – indicators of various sorts – without a good understanding of the dynamics of the system in a mechanistic sense, that is, why these correlations exist. If such correlations are established, management might be satisfactory even without much fundamental knowledge. For centuries before the advent of modern medicine, it was known that an elevated body temperature was a pathological sign and that certain treatments often alleviated the condition, but why body temperature was elevated was unknown. Even if such management procedures are successful for some communities and ecosystems, we should not minimize the huge amount of work that remains to be done for most systems to establish the correlations.

Second, a real understanding of the interactions within, and the dynamics of, communities and ecosystems might enable us to see clearly what factors contribute to biodiversity, and thus how to maintain biodiversity. Good management would then be predicated on accurate mechanistic models of the systems to be managed. Even if such models were a worthwhile scientific goal, however, it could be that the models available now are not sufficient to guide management and that indicators devoid of an underlying mechanism would be more useful for the present.

What Are Communities and Ecosystems?
Compared with individuals and populations, communities and ecosystems are very poorly understood (Simberloff 1990). For example, for almost all animal species there is no debate about where one individual ends and another begins; individuals have well-defined boundaries. To delineate populations is not so simple, for two reasons. First, we would have to determine which individuals mate with which other individuals. In practice, this determination is very difficult. But even if we were able to record all matings, we would still have problems delineating populations. Our problems occur because we must arbitrarily define the amount of gene flow that qualifies two groups of individuals to be considered as part of the same population. Recall that a population is traditionally defined as a group of individuals united by interbreeding, whereas two groups of conspecific individuals are defined as separate populations if gene flow between them is "infrequent." But exactly how little gene flow constitutes "infrequent" is arbitrary. There must be *some* gene flow or else the groups would not be conspecific. If we had perfect knowledge of who mates with whom, and if species were really structured as metapopulations, with discrete groups characterized by frequent interbreeding clearly separated spatially from one another and with much rarer interbreeding between groups, then delineation of populations would be straightforward. Most species in nature, however, do not seem to be arranged in this way (Harrison 1991; Simberloff 1993), so delimiting populations in nature is problematic.

Delimiting communities and ecosystems in nature is even more problematic. First, we have longstanding debate and a gradient of views about what communities are (Simberloff 1980; Underwood 1986). At one extreme, many ecologists have an "individualistic" conception of communities as simply the collection of populations present in one place at one time. In this conception, interactions among populations are not so numerous or stylized as to produce a highly organized entity. At the other extreme, some ecologists envisage holistic communities with emergent properties and with populations so intricately linked as to produce a superorganism (Wilson and Sober 1989).

A similar gradient characterizes views of the ecosystem, which is the community plus its physical environment. On the one hand, ecosystem functions such as energy flow and nutrient cycling can be seen as the inevitable trivial result of having several populations co-occurring. On the other hand, an ecosystem has been construed as a superorganism whose energy flow and nutrient cycling parallel an organism's physiology. These varying views of the nature of communities and ecosystems reflect the fact that we possess rather little general knowledge of how communities and ecosystems are organized (Paine 1988; Peters 1988). Because of this gap in knowledge, even simple perturbations of communities, such as the addition or deletion of a single species, seem to have idiosyncratic, unpredictable consequences that we do not understand and cannot predict (Simberloff 1995).

The physical boundaries of communities and ecosystems are often difficult to establish, and this difficulty is associated with the lack of consensus about what communities and ecosystems are. If they are highly organized, superorganismic entities, their spatial limits should be easily discerned. If, on the other hand, a community is simply a collection of populations, then it is not clear where one community ends and another begins, because each population has its own spatial limits and these need not coincide. Measuring the species diversity of communities would then be very problematic (Underwood 1986). Alpha diversity – the number of species within a community – would rest on what spatial boundaries are placed on the community. Beta diversity – the degree of difference between adjacent communities – would also change. The problem is that there would be no objective way to distinguish between a single community with very high alpha diversity and a group of communities, each with low alpha diversity but with great beta diversity.

Of course, even if one believes that the community is not a real or important level of ecological organization, one might concede, as a practical matter, a need to regulate and manage at the community level. For better or for worse, data exist on communities, and the public is interested in the trajectories and fates of communities. The need to manage communities does not necessarily mean, however, that one can do this best by focusing on community-level traits.

What Is Diversity at the Community and Ecosystem Level, and Why Are We Interested in It?

We measure diversity because we want to preserve it, but there are numerous kinds of diversity. The cost of measuring and preserving diversity can be enormous, so we must place priorities on our efforts. Thus we must consider carefully both what aspects of diversity are crucial to us and what methods are most effective to measure those aspects.

What Is Diversity?

"Diversity" at the community level originally simply meant number of species (species richness). Subsequently, we saw a flurry of interest in the idea that communities with the same numbers of species could be viewed as having different diversities depending on how individuals are apportioned among species. For example, consider two communities, each with 10 species and 100 individuals. If the first community has each of the 10 species represented by 10 individuals, while the second has 1 species represented by 91 individuals and the remaining 9 species represented by 1 individual each, in some intuitive sense the first community is more diverse than the second. Various statistics can be used to express diversity in this sense, which is often called "equitability" or "evenness." Particularly popular have been the information theoretic H' and evenness J, but a plethora of other statistics that account for varying degrees of numerical dominance among different species in a community all serve the same function of estimating equitability. In fact, almost all of these statistics can be viewed as special cases of a more general diversity function, with different values for certain parameters (Patil and Taillie 1979; Solomon 1979).

A complication arises when communities are compared using samples with different numbers of individuals (Magurran 1988; Schluter and Ricklefs 1993). The rarefaction procedure (Simberloff 1978), however, provides a simple, distribution-free, objective way to compare samples of different sizes. Rarefaction gives, for each sample, the numbers of species that might be expected, given certain assumptions, if all samples had the same numbers of individuals.

Numbers of species and the equitability of population sizes take no account of species identities, yet these identities contribute to diversity in some intuitive sense. For example, a plant community that consists of 100 species of congeneric grasses seems less diverse than one consisting of 100 species spread evenly among many different genera and families. Although this proposition seems self-evident, such "taxonomic diversity" (Simberloff 1970) has received little formal attention in terms of both how to quantify it and what to do about conserving it. Simple ratios such as mean number of species per genus or mean number of species per family have long been used by ecologists, evolutionists, and biogeographers for such comparisons (references in Simberloff 1970, 1978) and are valid so long as different sample sizes (different numbers of species) are taken into account. Again, rarefaction can be used for this purpose (Simberloff 1978). Of course, mean number of species per genus and related statistics would not always differentiate communities whose taxonomic diversities might seem intuitively very different. For example, a community with 10 genera, each with 10 species, would seem more diverse taxonomically than a community with 10 genera, 1 with 91 species and the other 9 with 1 species each, yet each would have a mean of 10 species per genus. We have no tradition, however, of expressing taxonomic diversity by using the various statistics described above that assess equitability of distributions.

More recently, Vane-Wright et al. (1991) have suggested cladistic systematics as a basis for measuring taxonomic diversity (which they call "taxic diversity"). The goal would be to conserve descendants of as many deep nodes in a cladogram as possible and to have the conserved deep nodes be as widely dispersed topologically on the cladogram as possible. One can easily understand the underlying rationale – to conserve the widest possible swath (in the phylogenetic sense) of diversity: birds and reptiles rather than birds and other birds. The success of such an enterprise, however, would be limited by how accurately the cladogram reflected evolutionary history, a question well beyond the scope of this paper. Suffice it to say that there is major controversy over whether current cladistic methods are likely to produce very accurate cladograms.

Franklin (1988) argues that the conception of diversity outlined above, resting essentially on how many and what kinds of species are present and how individuals are apportioned among species, is insufficient at the community and ecosystem levels. All of the above aspects of diversity relate to the species composition of a community and so might be termed "compositional diversity." Franklin (1988) would add process (or functional) and structural diversity as key elements of biodiversity. In forests, for example, he would tally large snags and fallen logs as elements of structural diversity, and nitrogen-fixing organisms as elements of functional diversity. The precise algorithms for tallying functional and structural diversity have not been proposed, although such venerable concepts as foliage height diversity (MacArthur 1964) may be guides. It appears, however, that Franklin's underlying reason for wanting to maintain such diversity is primarily because it will help maintain compositional diversity. A plethora of literature relates the presence of particular species to specific structural features such as snags, so it is quite reasonable to believe that maximizing structural diversity will maximize compositional diversity. Further, some aspects of structure might be easier to monitor than the species themselves.

Why Are We Interested in Diversity?

One often believes that all types of diversity – richness, equitability, taxonomic, functional, and structural – are desired for their own sakes. That is, one would want to conserve the most species, the most families or orders, etc. Many people see such a proposition as self-evident on aesthetic or moral grounds. But many attempts to demonstrate ethical and/or aesthetic requirements for conserving biodiversity (Norton 1987; Rolston 1991) are unconvincing. Others, such as the appeal for intergenerational equity enshrined in various international manifestos and protocols, may be intellectually compelling, but they have certainly not stopped the assault on biodiversity.

One may also wish, for ethical reasons, to conserve not all species in a particular reserve or forest but only species of particular concern, such as endemic species or endangered species. One may wish to eliminate introduced species. Or one may wish, for aesthetic reasons, to maintain species of particular visual or emotional appeal such as "charismatic mega-vertebrates." Again, such ethical and aesthetic concerns have not prevented massive habitat destruction and a great increase in extinction rate.

Thus we are driven to ask about the instrumental roles of biodiversity. How does biodiversity per se serve perceived and/or real human needs? The answer to this question determines which aspects of biodiversity are most important to target for conservation. Individual species themselves may be of importance to us. We can, for example, extract pharmaceuticals and other chemicals directly from individual species (Eisner 1991), or we may be able to extract genes from individual species that will help us to synthesize such chemicals. We may want to hunt individual species or to cultivate them as ornamentals.

On the other hand, we may be concerned with services provided by communities and ecosystems, such as water purification, flood control, and climate moderation (Ehrlich and Mooney 1983). If this were the case, biodiversity per se, in any of the senses described above, would be important only insofar as it contributed to ecosystem function. For example, the old saw that compositionally diverse communities and ecosystems are always more stable has finally faded as ecological dogma (Norton 1987; Shrader-Frechette and McCoy 1993); at least, it is now known to be an oversimplification. One might still argue, however, that more diverse ecosystems, on average, provide certain services better or for a longer time. For example, Franklin et al. (1989) contend that diversity in all the senses outlined above leads to greater productivity for forest systems. This assertion is unproven, but in any event we have a wealth of empirical evidence that many very simplified systems, such as agricultural or silvicultural monocultures, are prone to long-term effects of disasters such as pathogen and insect outbreaks.

Shortcuts and Indicators of Biodiversity

If one could measure and tabulate every feature of a community or eco-system, one could automatically determine the status and trend of any kind of biodiversity one cared about. One would have to identify the species, genetic constitution, physical location, and so on of every individual in the community. The size and complexity of natural communities make this task impossible. Thus we must seek statistics that describe biodiversity sufficiently well that we do not have to measure all of it. Summarizing any mass of data in a single statistic or index, or even a group of them, however, entails a loss of some of the original information. So it is important that the statistics and indices chosen represent adequately those elements of biodiversity that are most important to us.

Indicator Species

One possibility is that only certain species are important to us, for ethical, aesthetic, even instrumental reasons as suggested above, but we do not have enough information to know which ones they are. For example, there is currently tremendous interest in the possibility that genes from species in nature will be useful to humans, or at least will help some humans make money. The genes might be useful in their own right; for example, genetic engineers might put genes from a bacterium into a plant to confer resistance to an insect pest. Or a currently unused species might turn out to be valuable by virtue of its genes. For example, the Pacific yew (*Taxus brevifolia*), long considered virtually worthless compared with the major timber trees in Northwestern forests, is now valued because it contains taxol, a potent anticancer drug (Joyce 1993). If we are not confident of our ability to predict which species, or even which ecological or taxonomic groups of species, are likely to be shown valuable by future research, then all species must be the targets of our conservation efforts, and the best statistic or index would be the one most highly correlated with species richness. Such an index or statistic would also be the best one if we were truly interested in biodiversity per se rather than in some instrumental use of biodiversity.

Indicator species (Landres et al. 1988) have traditionally been used for this purpose, and indicators have been chosen whose presence is thought to reflect the greatest richness of species, at least of species in the target community, and whose population fluctuations are thought to reflect the fluctuations of the largest possible number of other species. Indicator species have also been used to assay chemical and physical environmental conditions, but we have no reason to believe that a species highly sensitive to a chemical pollutant, for example, thus qualifying as a good indicator species for this purpose, would reflect the presence and status of a particularly large number of other species. So one must be very clear about whether an indicator species is meant to indicate biodiversity.

If it is meant to indicate compositional biodiversity, it is not at all clear how such a species should be chosen (Landres et al. 1988). For example, Severinghaus (1981) suggested using the ecological guild concept on the grounds that the trajectory of one guild member should indicate the trajectories of others. On both empirical and theoretical grounds, however, this idea is highly questionable. Guild members have such different habitat requirements and other characteristics that one would not expect them to fluctuate in concert, and they do not (e.g., Block et al. 1986). Furthermore, competition among guild members might cause them to exclude one another, or at least to have complementary fluctuations in numbers. In fact, short of actually measuring both co-occurrence patterns and degree of correlation in temporal population fluctuations, I know of no proven way of selecting a biodiversity indicator from a group of candidate species. Some shortcuts, however, have been suggested.

One shortcut would be an indicator species explicitly chosen not to indicate compositional biodiversity directly but to indicate habitat quality, with the underlying assumption that habitat of high quality will support the most species, or at least the most species of interest. Probably any ecologist would agree with this idea as a general principle, at least for well-defined habitat types such as "old-growth longleaf pine forest" or "tall grass prairie." For any such habitat, certain specialist species are highly associated with the habitat and therefore with one another. Often these are the very species of concern, and although they may not be a majority of the species present even in prime stands of their favoured habitat, they are the ones whose status we would like an indicator species to indicate.

Two problems beset this approach in practice. First, various technical difficulties arise in estimating density and trends (Landres et al. 1988). Often these can be overcome. A more fundamental problem is controversy over what actually does constitute prime habitat for an indicator. For certain specialist species, such as those of many old-growth forests, most people would assume that the nature of prime habitat is so blindingly obvious that it does not even require precise empirical characterization. It is like pornography: "I can't define it exactly, but I know it when I see it." But the habitat requirements of what seems like a quintessential old-growth species, the Northern Spotted Owl (*Strix occidentalis caurina*), have been hotly contested. Many people (e.g., California Forestry Association 1993) have argued that this species is commonly found in a wide array of non–old-growth habitats. The Northern Spotted Owl has been designated a management indicator species for old-growth forest in the Northwest on the assumption that its strict habitat requirements would guarantee the presence of other, less well studied species (Lee 1985), such as amphibians (Welsh 1990) and specialist insects that appear to require old growth. If its habitat requirements are not, in fact, so strict, it clearly fails in this function. The Northern Spotted

Owl is, unfortunately, not a special case; the precise characterization of prime habitat for most species would probably be at least as controversial.

Kremen (1992) advocates using groups of indicator species chosen by ordination techniques. Such multivariate analyses allow a direct test of the extent to which the presence or density of particular species is correlated with presence or density of other species or with habitat variables. Interpretation of such exploratory analyses is often ambiguous, but the ecological significance of suggestive patterns can sometimes be tested by further field work. Another potential problem is that the number of significant axes extracted from an ordination may be so low that their characterization in habitat terms may be too imprecise and general to be useful. For example, location on a "moisture" axis or a "disturbance" axis may give insufficient information about a species' relationship with its habitat.

If we are interested in structural and functional biodiversity primarily as determinants of compositional biodiversity (see "What Is Diversity at the Community and Ecosystem Level, and Why Are We Interested in It?" above), a species or group of species whose presence and status correlate with those of many other species, just as we have been discussing, should suffice. If, however, we are interested in structural and functional biodiversity in their own right, such indicators may be inadequate. Noss (1990) suggests that, at the least, a group of species will have to be monitored, because no one of them will sufficiently correlate with enough other species and with other aspects of biodiversity. He further argues that such staples of academic community ecology as dominance-diversity curves, species richness and equitability measures, and proportions in different guilds will be needed to indicate changes in ecosystems and communities. For structural diversity he advocates the additional measurement of habitat variables often measured by wildlife biologists and foresters (such as canopy height and percent cover), and for functional diversity he proposes nutrient cycling rates, predation rates, and other standard measurements of ecologists. A problem here, of course, is that as the battery of indicators expands, it increasingly approximates the full set of traits to be indicated!

If our aim is to safeguard particular species – either those in need or charismatic ones – indicators of overall biodiversity will be of little use. Either the particular species themselves should be monitored, whether or not their vicissitudes correlate with those of other species, or habitat variables directly relevant to those species should be monitored. Some researchers think that particularly vulnerable species, especially habitat specialists, would be good indicators on the grounds that they would be especially sensitive to conditions that would subsequently affect many other species (references in Landres et al. 1988; see also Scientific Panel 1995). Although the traditional use of especially sensitive species as indicators of chemical pollution makes good sense (National Research Council 1986; Landres et al. 1988), we have no reason to believe that fluctuations of sensitive species correlate particularly well with those of a larger set of other species. Thus we have no reason to believe a priori that such species would be especially good indicators of biodiversity. Rather, their usefulness as indicators would have to be tested empirically, and if they are rare and/or declining, they might be poor indicators simply because they will be difficult to monitor.

Finally, if our interest in a community or ecosystem is purely instrumental and rests on services such as pollination or flood control that the system itself provides rather than on current or potential uses of particular species, very different indicator species, or suites of them, would be required. Or perhaps no indicator species at all would be useful. At least this would be the case if we had no evidence that the services rested on biodiversity itself. As noted above, Franklin et al. (1989) assert that biodiversity of itself increases productivity, but the authors present no evidence. The fact that stability and diversity are not inevitably monotonically related (references in Shrader-Frechette and McCoy 1993 and Tilman and Pacala 1993) makes it seem unlikely that diversity itself is always critical to any particular ecosystem function, but it might correlate with it. What is important to monitor, however, is the species or ecosystem traits that provide the services of interest.

If, for example, we were concerned with the climate control functions of a forest ecosystem, we would want to monitor species (the major canopy trees) that dominate evapotranspiration and any other species critical to them. The most efficient way to do this might be to monitor the desired service itself (in this case, evapotranspiration). Or it might be to monitor one or more of the species providing the service. But there would be no point in monitoring a species whose abundance was correlated with biodiversity itself.

We can, of course, give numerous practical reasons for choosing indicators, be they species or complex functions such as dominance-diversity curves. A species that fluctuates with many other species but that is itself almost impossible to study would be a poor indicator species. Equally problematic would be many of the ecological functions recommended by Noss (1990), which would require extensive sampling of many species in the community plus, in many instances, subsequent laboratory tests. The entire *raison d'être* of indicators is that, for logistic and economic reasons, it is impossible to monitor large portions of the community.

Keystone Species

Some predatory species, although they may be neither very numerous nor a conduit for a substantial fraction of the energy flowing through a community, can markedly change the composition and physical structure of a community. Paine (1966, 1969), studying marine invertebrates, called these "keystone" species, and found that at least one – the starfish *Pisaster ochraceus*

– prevents a mussel species that is its favourite prey from eliminating other species by outcompeting them for space. Similarly, rabbits in Britain play a keystone role in increasing the biodiversity of a community by preferentially grazing on grass species that would otherwise locally eliminate many other plants by competition for space (Harper 1969).

The keystone concept has been expanded. For example, Gilbert (1980) and Howe and Westley (1988) describe "keystone mutualists" as plant species that support many animal species that are themselves critical to other species (for example, as seed dispersers). The "umbrella" species (Shrader-Frechette and McCoy 1993) is similar to the keystone species in that, for some functional reason, it is responsible for the presence of other taxa (for example, by constituting or constructing their habitat). The difference is that umbrella species need not affect a substantial fraction of the entire community.

Mills et al. (1993) contend that the keystone concept, bastardized beyond all conservation utility because it has been applied to so many different kinds of species without very formal criteria, has become a panchreston. In particular, they argue that experimental removal and subsequent monitoring would be required to determine whether or not a species is really critical to the composition, structure, or function of a community. Without a formal definition and (usually) experimental verification of the importance of one species to other species, using keystone species as a conservation tool, they feel, is a bad idea. In fact, their critique resembles that of Landres et al. (1988) for indicator species. The absence of a clear definition hinders implementation, and, in any event, we might end up focusing on a species that is not really important. This problem would be exacerbated if our interest were in maintaining a particular species rather than all biodiversity (Mills et al. 1993) and if the particular species happened to be neither a keystone species nor closely associated with one.

But are Mills et al. (1993) not perhaps throwing out the baby with the bathwater (cf. Demaynadier and Hunter 1994)? After all, the fact that some, or even many, researchers have misused the term or have called a species a "keystone" without verifying its key role experimentally does not mean that the entire concept is invalid. The original experiments by Paine (1966, 1969) and those reported by Harper (1969) met the criteria and illuminated the process of maintaining compositional and structural biodiversity in particular communities. The real questions should be, do all communities have such keystone species, and if so, which are they? To the extent that a community has one or more keystones, monitoring their status and managing the system to maintain them should also maintain those features, such as compositional diversity, for which the link to the keystone species has been verified. The difference between a keystone species as indicator and an indicator species pure and simple is this: the fate of a well-chosen indicator species indicates the fate of other species, but we do not know why; all we know is that a correlation in presence-absence or numbers exists. The fate of a verified keystone species is tied to that of many other species because of particular activities of the keystone that we understand.

If one is interested solely in using an ecosystem to provide services as an ecosystem, a keystone species might be a good target to monitor as long as its persistence is truly critical to the continued integrity of the community. Probably the very concept of an ecosystem providing services *as an ecosystem* presupposes the view that the ecosystem is a highly integrated, organic entity. An advocate of the individualistic view would consider these services the products of particular species or small groups of them.

If one's concern with ecosystems and communities is for the services they provide, and particularly if one holds an individualistic view, much species loss and substituting of one species for another might be viewed as perfectly acceptable. Many of the papers in the compendium by Jordan et al. (1987) show that communities and ecosystems have been created bearing imperfect resemblance to any natural communities but still serving very well meteorological, hydrological, and other functional purposes (cf. Werner 1987; Cairns 1988). Again, the function of interest would likely be the thing to be monitored, and not a species that provides the function or "indicates" it.

Mechanistic Understanding and Resource Management

It is not clear that, from a practical standpoint, the understanding that accompanies using a keystone species rather than an indicator species is an advantage, however attractive it is from a scientific standpoint. If extensive empirical data show that an indicator's fate is correlated with the fate of many other species, even if we do not know why, is this fact not enough to establish that the indicator will be a useful tool?

This question is completely analogous to the debate that occasionally surfaces in ecology about whether prediction (for example, of fisheries or crop yields or pest numbers) suffices or whether a mechanistic model is better. For example, Poole (1974) and Peters (1991) argue that academic ecology is largely a failure in many applied areas because of reliance on unrealistic mechanistic models that do not provide accurate forecasts. Better, they argue, that one use simple statistical techniques (such as regression) with no underlying mechanistic model if the forecast is to be accurate.

Such views have raised a firestorm of protest from academic ecologists, especially modellers (Hengeveld 1992). In addition to being a reaction from practitioners of a method that is being attacked, these responses represent the philosophical stance that description and numerical prediction are simply lesser scientific achievements than is mechanistic understanding. Others, however, agree with Peters (1991) that mechanistic models have not, in fact, provided answers in many areas of applied ecology where problems are pressing and solutions, even more than mechanistic understanding, are urgently needed (Shrader-Frechette and McCoy 1993, and references therein). Bunnell (1989) has pointed to the same dichotomy in the purpose of models – prediction versus understanding – in forest wildlife management. Managers generally accept accurate prediction as sufficient, whereas scientists tend to view any model as more or less inaccurate but potentially useful in achieving mechanistic understanding.

Adaptive management (Walters 1986; Walters and Holling 1990) is similar to statistical forecasting in that it need not rely on a specific mechanistic model. Rather, the response of the system to some particular management scheme, whatever the underlying reason for the response, dictates the modification of the management scheme. Fishery management such as that of the North Pacific halibut fishery has been successful under this regime whereas mechanistic modelling failed dismally (National Research Council 1986). Adaptive management, explicitly recommended for managing national forests in the United States (Kessler et al. 1992), is part of a new emphasis on ecosystem management. Walters and Holling (1990) contend that one can learn scientifically from adaptive management. But the very fact that they argue that the knowledge they are talking about is science, even if it is not traditional, analytical scientific knowledge (which they call the "science of parts"), shows that they recognize the controversial nature of their contention. No one, however, would argue against their claim that adaptive management can produce effective and sustainable resource use, even if one might argue that the method does not promote mechanistic understanding of the system and that such understanding might lead to even more effective harvesting methods in the future.

Perhaps there is a middle ground (cf. Bunnell 1989; Hengeveld 1992). One can concede that scientists should aim for mechanistic understanding of a phenomenon, and even that mechanistic understanding would enhance management capabilities. But at the same time, one can admit that while our understanding of certain ecological phenomena, particularly at the community and ecosystem levels, is insufficient for management purposes, if accurate phenomenological techniques such as numerical prediction and use of indicator species can be devised, they ought at least to be used in the interim.

To an extent, the attractiveness of monitoring a keystone species as an indicator rests on one's conception of communities and ecosystems. If one views communities and ecosystems as holistic, organized, integrated entities with component species complexly intertwined, the role of a keystone species would be an expression of this organic nature. In fact, the existence of keystone species might constitute an emergent property, the Holy Grail of advocates of this concept of communities and ecosystems (Shrader-Frechette and McCoy 1993). It would make perfectly good sense to focus on the status and trends of such a keystone to understand the "health" of the community or ecosystem, in much the same way that it would make good sense to focus, for example, on the status and trends of the human heart or brain rather than on a finger or hair to estimate the health of a human being. It might turn out, of course, that the status of a finger or of hair is highly correlated, for unknown reasons, with the health of an entire human body. But it would still be peculiar to monitor fingers and hair rather than the heart unless, perhaps, it was much easier to monitor a finger or hair than to monitor the heart.

If, on the other hand, one conceives of a community as a collection of species that happen to be in the same place at the same time, not united by an intricate and stylized web of interactions, and the ecosystem as the physical matrix of that community, then the very existence of a keystone species might be suspect. If a keystone were convincingly demonstrated for such a community, it would be viewed as an oddity of that community, not an emergent property. It might be a good indicator species for the obvious reason that its fate is correlated with the fates of other species. But any species whose fate was equally correlated with those of many other species would be just as useful an indicator because we are basically interested in the fates of all the other species, not so much of the community per se composed of these species.

Acknowledgments

I thank Michael Beck and Fred Bunnell for numerous suggestions on the manuscript.

Literature Cited

Block, W.M., L.A. Brennan, and R.J. Gutierrez. 1986. The use of guilds and guild-indicator species for assessing habitat suitability. Pp. 109-13 in J. Verner, M.L. Morrison, and C.J. Ralph (eds.). Wildlife 2000: Modeling habitat relationships of terrestrial vertebrates. University of Wisconsin Press, Madison, WI.

Bunnell, F.L. 1989. Alchemy and uncertainty: What good are models? General Technical Report PNW-GTR-232, US Department of Agriculture (USDA) Forest Service, Portland, OR.

Cairns, J. 1988. Rehabilitating damaged ecosystems, Vols. 1 and 2. CRC Press, Boca Raton, FL.

California Forestry Association. 1993. A petition to remove the California population of the Northern Spotted Owl from the federal list of threatened species. California Forestry Association, Sacramento, CA.

Demaynadier, P., and M.L. Hunter. 1994. Keystone support (letter). BioScience 44:2.

Ehrlich, P.R., and H.A. Mooney. 1983. Extinction, substitution, and ecosystem services. BioScience 33:248-54.

Eisner, T. 1991. Chemical prospecting: A proposal for action. Pp. 196-202 in F.H. Bormann and S.R. Kellert (eds.). Ecology, economics, ethics: The broken circle. Yale University Press, New Haven, CT.

Franklin, J.F. 1988. Structural and functional diversity in temperate forests. Pp. 166-75 in E.O. Wilson (ed.). Biodiversity. National Academy Press, Washington, DC.

Franklin, J.F., D.A. Perry, T.D. Schowalter, M.E. Harmon, A. McKee, and T.A. Spies. 1989. Importance of ecological diversity in maintaining long-term site productivity. Pp. 83-96 in D.A. Perry, R. Meurisse, B. Thomas, R. Miller, J. Boyle, J. Means, C.R. Perry, and R.F.

Powers (eds.). Maintaining the long-term productivity of Pacific Northwest forest ecosystems. Timber Press, Portland, OR.

Gilbert, L.E. 1980. Food web organization and the conservation of neotropical biodiversity. Pp. 11-33 *in* M.E. Soulé and B.A. Wilcox (eds.). Conservation biology: An evolutionary-ecological perspective. Sinauer Associates, Sunderland, MA.

Harper, J.L. 1969. The role of predation in vegetational diversity. Pp. 48-62 *in* G.M. Woodwell and H.H. Smith (eds.). Diversity and stability in ecological systems, Brookhaven Symposia in Biology Number 22. Brookhaven National Laboratory, Upton, NY.

Harrison, S. 1991. Local extinction in a metapopulation context: An empirical evaluation. Biological Journal of the Linnean Society 42:73-88.

Hengeveld, R. 1992. Right and wrong in ecological explanation. Journal of Biogeography 19:345-47.

Howe, H.F., and L.C. Westley. 1988. Ecological relationships of plants and animals. Oxford University Press, New York, NY.

Jordan, W.R., III, M.E. Gilpin, and J.D. Aber (eds.). 1987. Restoration ecology: A synthetic approach to ecological research. Cambridge University Press, Cambridge, UK.

Joyce, C. 1993. Taxol: Search for a cancer drug. BioScience 43:133-36.

Kessler, W.R., H. Salwasser, C.W. Cartwright, Jr., and J.A. Caplan. 1992. New perspectives for sustainable natural resources management. Ecological Applications 2:221-25.

Kremen, C. 1992. Assessing the indicator properties of species assemblages for natural areas monitoring. Ecological Applications 2:203-17.

Landres, P.B., J. Verner, and J.W. Thomas. 1988. Ecological uses of vertebrate indicator species: A critique. Conservation Biology 2:316-28.

Lee, P.L. 1985. History and current status of Spotted Owl (*Strix occidentalis*) habitat management in the Pacific Northwest Region, USDA Forest Service. Pp. 5-9 *in* R.J. Gutierrez and A.B. Carey (eds.). Ecology and management of the Spotted Owl in the Pacific Northwest. Technical Report PNW-185, US Forest Service, Portland, OR.

MacArthur, R.H. 1964. Environmental factors affecting bird species diversity. American Naturalist 98:387-97.

Magurran, A.E. 1988. Ecological diversity and its measurement. Princeton University Press, Princeton, NJ.

Mills, L.S., M.E. Soulé, and D.F. Doak. 1993. The keystone-species concept in ecology and conservation. BioScience 43:219-24.

National Research Council Committee on the Applications of Ecological Theory to Environmental Problems. 1986. Ecological knowledge and environmental problem-solving. National Academy Press, Washington, DC.

Norton, B.G. 1987. Why preserve natural variety? Princeton University Press, Princeton, NJ.

Noss, R.F. 1990. Indicators for monitoring biodiversity: A hierarchical approach. Conservation Biology 4:355-64.

Paine, R.T. 1966. Food web complexity and species diversity. American Naturalist 100:65-75.

–. 1969. A note on trophic complexity and community stability. American Naturalist 103:91-93.

–. 1988. Food webs: Road maps of interactions or grist for theoretical development? Ecology 69:1648-54.

Patil, G.P., and C. Taillie. 1979. An overview of diversity. Pp. 3-27 *in* J.F. Grassle, G.P. Patil, W. Smith, and C. Taillie (eds.). Ecological diversity in theory and practice. International Co-operative Publishing House, Fairland, MD.

Peters, R.H. 1988. Some general problems for ecology illustrated by food web theory. Ecology 69:1673-76.

–. 1991. A critique for ecology. Cambridge University Press, Cambridge, UK.

Poole, R.W. 1974. An introduction to quantitative ecology. McGraw-Hill, New York, NY.

Rolston, H., III. 1991. Environmental ethics: Values in and duties to the natural world. Pp. 73-96 *in* F.H. Bormann and S.R. Kellert (eds.). Ecology, economics, ethics: The broken circle. Yale University Press, New Haven, CT.

Schluter, D., and R.E. Ricklefs. 1993. Species diversity: An introduction to the problem. Pp. 1-10 *in* R.E. Ricklefs and D. Schluter (eds.). Species diversity in ecological communities: Historical and geographic perspectives. University of Chicago Press, Chicago, IL.

Scientific Panel for Sustainable Forest Practices in Clayoquot Sound. 1995. Report 5, Sustainable ecosystem management in Clayoquot Sound: Planning and practices. Ministry of Environment, Lands and Parks, Victoria, BC.

Severinghaus, W.D. 1981. Guild theory development as a mechanism for assessing environmental impact. Environmental Management 5:187-90.

Shrader-Frechette, K.S., and E.D. McCoy. 1993. Methods in ecology: Strategies for conservation. Cambridge University Press, Cambridge, UK.

Simberloff, D. 1970. Taxonomic diversity of island biotas. Evolution 24:23-47.

–. 1978. Use of rarefaction and related methods in ecology. Pp. 150-65 *in* K.L. Dickson, J. Cairns, Jr., and R.J. Livingston (eds.). Biological data in water pollution assessment: Quantitative and statistical analyses. American Society for Testing and Materials, Philadelphia, PA.

–. 1980. A succession of paradigms in ecology: Essentialism to materialism and probabilism. Synthese 43:3-39.

–. 1990. Reconstructing the ambiguous: Can island ecosystems be restored? Pp. 37-51 *in* D.R. Towns, C.H. Daugherty, and I.A.E. Atkinson (eds.). Ecological restoration of New Zealand Islands. Department of Conservation, Wellington, NZ.

–. 1993. The ecology of extinction. Acta Palaeontologica Polonica 38:159-74.

–. 1995. Introduced species. Pp. 323-36 *in* W.A. Nierenberg (ed.). Encyclopedia of environmental biology, Vol. 2. Academic Press, San Diego, CA.

Solomon, D.L. 1979. A comparative approach to species diversity. Pp. 29-35 *in* J.F. Grassle, G.P. Patil, W. Smith, and C. Taillie (eds.). Ecological diversity in theory and practice. International Co-operative Publishing House, Fairland, MD.

Tilman, D., and S. Pacala. 1993. The maintenance of species richness in plant communities. Pp. 13-25 *in* R.E. Ricklefs and D. Schluter (eds.). Species diversity in ecological communities: Historical and geographic perspectives. University of Chicago Press, Chicago, IL.

Underwood, A.J. 1986. What is a community? Pp. 351-67 *in* D.M. Raup and D. Jablonski (eds.). Patterns and processes in the history of life. Springer-Verlag, Berlin.

Vane-Wright, R.I., C.J. Humphries, and P.H. Williams. 1991. What to protect? – Systematics and the agonies of choice. Biological Conservation 55:235-54.

Walters, C.J. 1986. Adaptive management of natural resources. McGraw-Hill, New York, NY.

Walters, C.J., and C.S. Holling. 1990. Large-scale management experiments and learning by doing. Ecology 71:2060-68.

Welsh, H.H., Jr. 1990. Relictual amphibians and old growth forests. Conservation Biology 4:309-19.

Werner, P. 1987. Reflections on "mechanistic" experiments in ecological restoration. Pp. 321-28 *in* W.R. Jordan, III, M.E. Gilpin, and J.D. Aber (eds.). Restoration ecology: A synthetic approach to ecological research. Cambridge University Press, Cambridge, UK.

Wilson, D.S., and E. Sober. 1989. Reviving the superorganism. Journal of Theoretical Biology 136:337-56.

6
Biodiversity at the Landscape Level
J. Stan Rowe

Introduction

"Scientists have a very rudimentary knowledge of biodiversity," said Solbrig in 1991, a statement whose accuracy will stand unchallenged for many years. Worldwide, only about 1.4 million species of organisms out of a probable 30 or 40 million have been described. In Canada, Mosquin et al. (1994) place the number of identified species – the big ones – at around 70,000, which means that many thousand little ones are lurking incognito in the northern mountains, forests, grasslands, and tundra. Clearly the biodiversity of any area depends on its "taxodiversity": the number of different kinds of taxonomists that have packed or been packed into it. Because we know that organisms are inseparable from their landscape or waterscape systems, and because identifying different kinds of land or water units is relatively easy (compared with cataloguing organisms), a preservation program for terrain in its inclusive sense can go a long way towards compensating for taxonomic ignorance. Further, a landscape-ecosystem paradigm for conservation is an improvement over the narrow traditional, species-oriented paradigm. Species conservation, undoubtedly important, draws funding because it is the focus of public interest. Our educational system has not taught the inherent value of the systems within which species dwell.

Previous discussion in this volume has been about biodiversity at the community-ecosystem level. When biodiversity at the community-ecosystem level is placed in the context of landscape, where "environment" takes on the reality of landforms, soils, water bodies, and climate, then much of the impermanence and indiscreteness of the community-ecosystem concept falls away. Organisms within the matrix of the landscape mosaic are perceived as dependent though contributing parts of larger entities. Ecosystem wholes emerge as the important entities, with communities as subsets of them – a subtle reversal of the lower-level terminology.

Perhaps "biodiversity" should always appear in quotation marks to signify an indecisive term. As yet we have no comprehensive, rigorous, and intellectually satisfying concept of what it is. Different strands of ideas have been wound into it at different times, some, such as "ecosystem," as after-thoughts. But because organisms cannot long survive without their equally creative matrix of air, water, and soil, preserving ecosystems at the land-scape level is the necessary practical approach for all those in land manage-ment concerned with biodiversity. This is the "filter" that, if made sufficiently large, catches everything, whether we know it on sight or not. Safeguarding biodiversity comes down to enlightened land/water preservation and man-agement in the context of long-range, comprehensive land-use planning. With the exception of forlorn specimens in zoos, botanical gardens, and other ex situ arks, organic forms sink or swim according to the fate of the landscapes or waterscapes that gave birth to them and now support them.

On Saving Life

We have tackled the problem of preserving "life" from the wrong end any-way. In the past, ecological intuition attributed "life" not only to organisms but also to what surrounds them: to earth, to water, to the *anima* and *animus* of the air – and why not? Whatever that mysterious organizing principle called "life" may be, its immediate source is clearly the ecosphere. Ecology demonstrates that organisms and their earthly matrices are parts of an in-separable continuum, differentiated only by our cheating senses. A creative animating process, life is an expression of the Blue Planet and its 4.6 billion years of evolution. The Biological Fallacy equating organisms with life is the result of a faulty inside-the-system view (Rowe 1992a).

Consider the different experiences of seeing a system from the outside and from the inside. Looking through a microscope at a slice of plant tissue, a student sees spaces bounded by walls and knows, from the instructor, that he or she is seeing unitary things called cells. Next, looking within, the student is mentally prepared to see parts: nuclei, plastids, mitochondria, starch grains, streaming cytoplasm, particles dancing in Brownian move-ment. Note that the identification of parts is contingent on prior definition of the whole, as shown by a simple thought experiment.

Suppose that instructor and student, before seeing cells from the outside, were reduced to microscopic size and placed within a cell. The teacher hands binoculars to the student and asks, "What do you see?" Sight from within particularizes; lacking the outside perspective that reveals the whole, the student will see the cell contents as separate and unconnected objects. He or she might then logically identify the dividing, reproducing organelles as alive and their cytoplasmic matrix, vacuoles, and plasma membrane as dead. The idea that the totality is alive, so obvious from the outside, is not appar-ent from inside.

For thousands of years humans have been viewers immersed in the ecosphere, deep-air animals living at the phase boundaries where air and water meet land, narrowly classifying all manner of things as organic and inorganic, biotic and abiotic, animate and inanimate, living and dead. Dictionaries full of nouns show the efficiency with which we have thought the world to pieces. Around our ignorant taxonomy we have constructed religions, philosophies, and sciences that fragment and compartmentalize a global ecosystem whose "aliveness" is expressed as much in its improbable atmosphere, crustal rocks, seas, soils, and sediments as in organisms. When did life begin? When the ecosphere itself was born, if not even earlier.

When life is conceived as a property of the ecosphere and its sectoral ecosystems, the subject matter of biology is cast in a bright new light. Earth spaces begin to look as important as earth species. When animation is perceived where it belongs, the focus shifts to ecological diversity, the diversity, to use Leopold's phrase, of "land communities," which are the creative entities to which organisms belong and from which they came. Without this broader concept, the tendency is to concentrate too narrowly on "saving organisms and their habitats," whose final solution, in the minds of some urban-dwellers, is bigger zoos and more wildlife ranches.

Regionalizing the Land

The multiple scales of intrusive human uses in space and over time, particularly in agriculture and forestry, are the scales at which diversity is threatened. Therefore the land (the term includes its water bodies) is best comprehended in terms of units of different sizes or scales. Most efficient and most easily understood is a hierarchical system of land units such as the ones soil surveys have provided for agriculture and those that forest site regionalizations and classifications are providing for forestry. The importance of "ecological land classifications" is indicated by its prominence in the National Forest Policy (Canadian Council of Forest Ministers 1992), where the first commitment reads: "Governments will complete an ecological classification of forest lands." This statement expresses the general recognition that current problems of conservation having to do with biodiversity, old growth, wildlife habitat, water, and recreational land use must be planned spatially in the forest domain with at least a modicum of understanding of the regional terrain mosaic.

In British Columbia, foresters use the Biogeoclimatic Ecosystem Classification system (BEC system) to stratify the land into large biogeoclimatic zones expressing major forest-climate divisions (MacKinnon et al. 1992), which are then further divided into subzones according to climate and vegetation. Subzones are composed of "site series": topographic or landform sequences of units that are each about "stand" size. Each site series unit or portion of the landscape, including the life on and in it, is a biogeocoenose or an ecosystem (Meidinger and Pojar 1991). Field guides have been prepared to assist the forester in identifying forest stands by subzone and site series as well as providing advice on such silvicultural activities as matching species to site for planting programs and assessing productivity. Because each unit in the hierarchy is mappable, the BEC provides a framework for a variety of land-use decisions, including preserving biodiversity.

To complement this, the British Columbia Ministry of Environment, Lands and Parks has devised a scheme that divides the province into 10 ecoprovinces, 30 ecoregions, and 68 ecosections (Demarchi et al. 1990). Similar systems of land regionalization are in use or being developed in all parts of the country; their historical roots go back 60 years (Sims 1992). Using both physiographical and biological criteria, Environment Canada has mapped the country into 15 ecozones, 40 ecoprovinces, 177 ecoregions, and 5,400 ecodistricts in a hierarchical scheme whose chief virtue is that the units are tied to map scales. For example, the ecoregions used as the basis for a cross-country check-up on biodiversity by Mosquin et al. (1994) are scaled from 1:1,000,000 to 1:3,000,000. The smaller "ecosection" – a common unit in forest regionalization – is scaled at 1:50,000 to 1:250,000 (Ironside 1991).

Despite systems of regionalization used by agriculture and forestry for many decades, management of biodiversity at the landscape level is hampered by lack of agreement on the definition of a landscape unit. Two possible reasons are: (1) disagreement about the criteria for landscape units, and (2) unfamiliarity with the earth-science approach to defining such landscape units as terrestrial ecosystems, forest sites, and soils.

Criteria for Defining Landscape Units

To explore this problem, consider the letter by aggrieved members of the US Soil Ecology Society stating their conviction that "the belowground component is essential for all terrestrial ecosystems" and asking why three-dimensional soil systems have not been fully addressed in the Ecological Society of America's Sustainable Biosphere Initiative (Klopatek et al. 1992). The answer appears to be that earth-science ecologists and biological ecologists are on different conceptual tracks when it comes to the meaning of "ecosystem" (Rowe and Barnes 1994). Most bio-ecologists centre their ecosystems on organisms, whether mobile (animal ecologists) or relatively sedentary (plant ecologists), adding thereto a nonspecific "abiotic" environment. "The sub-entities that make up an ecosystem are the biotic community *and the abiotic resources*" (emphasis added), according to Scheiner et al. (1993). The bio-ecologist's ecosystem tends to be a conceptual device relatively elastic in its space/time dimensions, deriving its meaning or lack of meaning from organisms of interest wherever they chance to roam. "I have struggled unsuccessfully with the problem of defining ecosystems into which a seagull can be fitted," said Drury in 1969, and similar sentiments have been expressed ever since.

Geo-ecologists have no such problem. Their ecosystem resembles a giant terrarium or aquarium with a particular developmental history at a particular location (Barnes 1993). It is a volumetric, layered, site-specific object such as a lake, a particular landform-based forest, or a more complex tract of land and water terrain into and out of which mobile organisms come and go. Where the bio-ecologist's ecosystem tends to be abstract and heuristic, its purpose being to add an "abiotic" dimension to studies of individual, population, and community, the geo-ecologist's ecosystem is a real live chunk of earth space, less abstract than the taxonomic categories of population/community and with the same spatial/structural concreteness (once boundaries have been set) as an individual organism (Rowe 1992b).

The two ideas concealed within the term *ecosystem* can be explained as arising from complementary ways of knowing, that is, as epistemological. Bio-ecologists have come to the concept "from below," *by addition*, starting with organisms and then adding (under the category "environment") whatever other aspects of reality seemed relevant. The ontogeny of the concept is illustrated in ecological textbooks that begin with the individual (autecology), then progress by addition to population ecology followed by community ecology, and then, at the end, the ecosystem concept (commonly defined as "community plus physical environment"). Note in these examples that the objects of primary interest are *organisms*, not soils or lakes or coral reefs or deserts or tracts of forestland.

The second thought process, illustrated by the approach geo-ecologists take to many of their field problems, proceeds "from above," *by division*. The geo-ecologist begins with some large natural system of land/air/water/ organisms, and by subdivision arrives at component ecosystems whose scale or size depends on the inherent diversity of the terrain that is the focus of human purpose. Thus the field pedologist, working in a regional or zonal context, recognizes and maps three-dimensional soils or soil complexes, "the belowground component ... essential for all terrestrial ecosystems" (Klopatek et al. 1992). While soil scientists recognize the importance of organisms in soil genesis (the biota being one of five relatively independent variables), their primary interest is in the volumetric system rather than in organisms per se (Jenny 1980). To them the matrix that surrounds soil organisms is not so much an "environment" (namely, a matrix defined by its importance to organisms) as a three-dimensional "natural body," a supraorganic ecosystem (or the basal component of an ecosystem) within which organisms function as vital components. They therefore set ecosystem boundaries according to multiple criteria – often using visible ecological relationships between vegetation and landform – rather than according to the single biological criterion of presence or absence of particular organisms.

The logic of addition and the logic of division generate different meanings under the same name. The second and more holistic idea of "ecosystem" is especially popular outside academe, where resource managers of all stripes now proclaim their goal as ecosystem management, implying that particular geographic land/water areas or landscape units plus all that is within them are the ecosystems of interest. For those interested in landscape-level resource planning, this approach makes sense. In contrast, conceptualizing "ecosystem" by adding abiotic resources to a dealer's choice of organisms reduces the term to the same vague and subjective status as "environment." By this route every organism defines its own ecosystem, and ecosystems are infinite in number, non-coincident in boundaries, and virtually inaccessible to study. No wonder the "ecosystem or landscape approach" has been said to lack scientific rigour. In the following discussion the term *ecosystem* is used in the place-specific earth-science sense, whose definition and location geographically can be as rational and rigorous as the units studied in geomorphology, soil science, and topoclimatology.

An Earth-Science Approach to Defining Landscape Units
It is helpful to think of ecosystems as chunks of earth space of all sizes, forming a nested hierarchy that extends from the ecosphere and its major subdivisions down to smaller waterscapes and landscapes, named on a descending size scale such as ecozones, ecoprovinces, ecoregions, and so on, to the level of small, local air/landform/soil/organism volumes. Every ecosystem unit, large or small (at least below the ecoregional level) can also be called a terrestrial landscape ecosystem or an aquatic waterscape ecosystem, and each is in intimate and dependent contact with the larger surrounding region of which it is a part. Thus regions and landscapes are also ecosystems, but using these terms necessarily implies awareness of size and scale. To plan and practice forestry "at the landscape level" is vague until the size and scale are specified. In British Columbia, the "landscape level" of forestry planning pertains to "a watershed, or series of interacting watersheds or other natural biophysical (ecological) units" within a subregion, the areas in mind being 10,000 to 100,000 ha in size (BC Forest Service and BC Environment 1995) – on the smallish side for the kind of planning needed.
Several important considerations include the following:

• Units of the landscape are conceived as three-dimensional entities. Like lakes, the landforms, with soils and climates that cling to their surfaces, are place-specific and thus amenable to mapping, to study on the ground, and to revisiting more than once.
• The units are two-layered, air over land/water, with organisms clustered near the phase boundaries. Rarely has the air layer been considered a vital component, presumably because neither the Bible's Genesis story of creation mentions it nor did Aldo Leopold when he defined what he meant by the "biotic community" or "land community." Yet the air layer, the

carrier of essential gases and nutrients and the medium wherein climate is most vividly expressed, merits inclusion as a vital component of landscape/waterscape ecosystems, as do the landforms and waterforms that exert modifying effects on the air layer and its climate near the ground. Like soil diversity, whose preservation ought also to be a focus of interest, the diversity of local climates and of organisms is a reflection of ecodiversity and will fade away if attention is not given to maintaining ecosystems as wholes.

- The units of various sizes, nested hierarchically, are "natural bodies," "land," or "landscapes"; the geographer Carl Troll studied them, beginning in the 1930s, under the name "Landschaftsökologie," translated as "Landscape Ecology." In examining early aerial photographs, Troll was impressed by the ecological interconnections between local landscapes within their larger regions such as watersheds. For example, adjacent slopes affect one another, river valley landforms develop as integrated parts of the whole, the characteristics of wetlands are strongly influenced by the landscapes that surround them, and animals find their various wants satisfied in patchy landscapes. Ecological connectivity characterizes each natural region, joining by process and function the landscapes and waterscapes that the region comprises. Troll (1971) pointed out that these "units of nature" can be studied physiologically (processes within) and ecologically (relationships to what is without).

- The diversity of the units at any level within units of the higher levels of the hierarchy is a diversity of ecosystems, or "ecodiversity," amenable to measurement as alpha and beta diversity and providing another useful measure of biodiversity. As pointed out by Boyle (1992), quantifying habitat or ecosystem diversity has received far less attention than species or genetic diversity.

- Landform is the anchor and control of terrestrial ecosystems. It is the feature that contributes most to their unique ecological relationships and mappability. As a key point, this deserves a full explanation.

The Importance of Landform to "Landscape-Level" Biodiversity

Naturalists know that to find different plants and animals they must go to different kinds of landforms: to sand plains; to hummocky moraine; to rock outcrops; to bogs and fens; to hills, cliffs, valleys; to north-facing and south-facing slopes; to the full range of landform/soil/local climate sites. Furthermore, they know that on these sites they will find different flora and fauna according to the stage of succession (time since disturbance) as initiated and influenced by fire, wind, flood, pathogens, and so on. Thus ecosystem diversity is the expression of different disturbance regimes and histories superimposed on the diversity of terrain.

As defined earlier, landscape ecosystems are dynamic three-dimensional "boxes" consisting of an atmospheric stratum overlying a soil/water stratum with organisms clustered at the phase boundaries. Soil scientist Hans Jenny (1941) made the argument for five relatively independent "genetic" influences on soil formation: relief/topography, geological parent material, climate, biota, and time. The same "genetic" influences shape landscape ecosystems. If within a region any four of the five are held constant, variations in the fifth will be faithfully reflected in landscape or soil variability. Jenny's formulation can be simplified to four genetic influences by combining relief/topography and geological parent material as *landform* (defined as surface relief/topography as well as composition of the surficial parent material), along with *climate, biota,* and *time.* Then within any moderately uniform climatic region (macro, meso, or micro), given an equivalent time span and equal access overall to the same biota, all major variations in landscape ecosystems can initially and fundamentally be attributed to variations in landforms as they modify radiation and moisture regimes, select the fit plants and animals from the available biota, and modify the formation of soils on their topographic facets.

In mountainous terrain, vastly different climates can appear within relatively small vertical and horizontal distances, hence the landform/climate relationship is exceedingly complex. Nevertheless, landform at all lesser scales or sizes within ecological zones, subzones, and regions is the main contributor to and controller of both nutrient regime and climatic regime (radiation and moisture) near the ground, the source of the edatopic grids used to characterize site units in the biogeoclimatic (land) classification system of the BC Forest Service (Meidinger and Pojar 1991). Thus, in effect each different kind of landform engenders vital characteristics in the ecosystems or seral sequence of ecosystems (chronosequence) that encompass it.

Although strongly responding to and reflecting the influences of landform, the biota are also influenced by ongoing landscape processes. Therefore ecosystem diversity is the expression of different disturbance regimes and processes superimposed on physiographic diversity. This means that protection of forest biodiversity requires protection of the entire suite of forest types in all their successional stages, from youth to old age, on each different kind of landform within each major region: on colluvial slopes, on till uplands, on outwash plains, on lacustrine and alluvial flats, on sand dunes, on organic soils. By identifying "enduring features of the landscape" (parent material origin and texture, and relief/topography), the World Wildlife Fund's Protected Areas Gap Analysis Methodology seeks to identify sources of ecological and biological diversity within the country's natural regions, that is, at the ecoregion level (Kavanagh and Iacobelli 1995). At larger scales of 1:50,000 to 1:15,000, mapping of landscape ecosystems is feasible, as shown by the biophysical surveys of the National Parks (for example, Achuff et al. 1984) and by the work of Barnes and his associates in Michigan (Albert et al. 1986; Barnes 1993).

An important implication of the emphasis on ecodiversity is that the earth sciences – pedology, geology, geomorphology, hydrology, topoclimatology – are recognized as the equals of the biological sciences. Interdisciplinary research, study, and management will involve many more players than in the past. Those trained as biologists, pedologists, or hydrologists are in fact experts in subdisciplines of ecosystemology. The most important aspects of ecosystem ministration will often pertain to preserving landforms, soils, and water regimes in the landscape, these being the "enduring features" that are the prerequisites of ecosystem functions, which, in turn, support the taxonomic diversity of organisms.

Because of the periodic disturbance and dynamism of natural ecosystems, natural ecosystems can be preserved only as parts of regional complexes with large protected nodes and connecting corridors (particularly riparian corridors) that allow for back-and-forth migrations and gene flows. The ideal, if ever it can be attained, is the "Swiss cheese model" (Hammond 1991), where the holes are the nodes of human activity embedded healthfully in the larger nature-matrix. This idea implies land-use planning on a grand scale and with a long time horizon. Maintaining natural landscape/water-scape ecosystems is possible only within a scheme of ecological land-use planning that looks far into the future. Today's primary question might be phrased this way: what should Earth's landscapes and waterscapes be – structurally, compositionally, functionally, and aesthetically – 100, 500, 1,000 years from now? The many publications of Reed Noss (e.g., 1990, 1992a, 1992b) outline the problems and future opportunities for conserving and restoring biodiversity in large ecoregions and bioregions.

Comprehensive Ecological Land-Use Planning
Pushing diversity aside, humanity steadily pursues its own exclusive monoculture while paying lip service to preserving the ecosphere. To slow the first trend while making a reality of the second requires the strict preservation of large tracts of the earth's surface. How much is enough?

Odum (1970) suggested that 40% of his home state of Georgia should be preserved in the natural condition, with 30% allotted to food production, 20% to fibre production, and 10% to urban-industrial systems. This could be done, he said, without affecting the standard of living. Odum and Odum (1972) proposed a 50/50 split between natural lands and cultured lands for southern Florida.

Through numerous studies, Noss (1992a, 1992b, and Chapter 7 of this volume) also arrived at the figure of 50% of land/water in a natural state as "a credible estimate of what it takes to maintain viable populations of large

carnivores and to represent at least some landscapes where natural disturbance regimes can operate unimpeded." Noss has taken the bull by the
horns and set down his ideas in maps at the scale of the bioregion, blocking
out the primary and secondary reserves, the protective multiple-use buffer
zones, and the corridors and landscape connections. Perhaps the best way
to represent all ecosystems is to maintain the full array of physical habitats
and environmental gradients in reserves, from the highest to the lowest
elevations, from the driest to the wettest sites, and across all types of
substrates and topoclimates.

Mosquin et al. (1994) adopted the "ideal balance" of land uses envisaged
by Naess (1989), recommending "that the provincial and federal governments cooperate to develop land-use legislation, policies and zoning aimed
at the ultimate goal of allocating about one-third of the lands and water in
each ecozone ... for the purpose of retaining and/or restoring wild species
and natural ecosystems where the laws of nature would operate (as they
have in the past) without major human intrusions." Nothing less, they argue, can maintain the vital "ecological functions" of the world. They further propose that one-third of the lands and water be zoned to meet
humanity's essential productive and living needs and one-third be zoned
for sustainable extraction of biological resources.

At the moment, the popular figure for land/water preservation in Canada
is 12% by the year 2000, a target sponsored by the World Wildlife Fund and
endorsed by provincial and territorial governments. Ten to 12% preservation as Forest Ecosystem Networks for each site series has been recommended
in the working coastal forests of British Columbia (BC Forest Service and BC
Environment 1995). That this figure should in some quarters be viewed as
excessive is explained by the fact that in 1994 only 2 or 3% of the country's
terrestrial ecosystems lay within fully protected areas (currently 5.7%). An
expansion of wildernesses, parks, and other strict preservation areas to meet
the 12% total means subtracting 9 or 10% from exploitable lands.

The 12% target has been interpreted in some quarters as the ransom paid
to set industry free to carry on as before. The idea is to zone out 12% and
devote the other 88% to multiple-use and single-industrial ends, a "triad"
of forest land allocations (Seymour and Hunter 1992; Blouin 1993; Canadian Pulp and Paper Association 1993). According to this logic, acceptance
by provincial and territorial governments of 12% as the goal for "full-
protection forests" leaves 88% available for exploitation in various traditional ways. But this will not protect ecodiversity and biodiversity. The
alternative in the forest domain is a system of protected forest ecosystems
plus reform of practices outside them so as to maintain all ecosystem values
(Burton et al. 1992). As long as the traditional industrial model of forestry is
accepted as valid and perpetuated on a large fraction of forestland, its values will continue to be the norm, obscuring the need for a paradigm shift.

In those semi-natural ecosystems where extractive pursuits such as logging are practised, management practices should heed the historical patterns (spatial and temporal) of wild ecosystems, thus giving rare animals and plants a chance to save themselves. The only way to satisfy the popular demand for preserving biodiversity is to practise silviculture and harvesting within large regions in ways that maintain landscape ecosystems in mosaic patterns that approximate or mimic natural mosaic patterns.[1] For example, in lightning- or fire-induced boreal forests the pattern seems to be a few very large, irregular patches containing unburned inclusions; middling numbers of intermediate-sized irregular patches; and a very large number of small irregular strips and patches. As to age-class distribution in the boreal landscape, studies indicate that it roughly fits a negative exponential curve – that is, a great many young stands and an intermediate number of middle-aged stands tailing off to mature and overmature stands occupying about 30% of the area. With an average fire rotation of 100 years, this means that the normal distribution of forest ages will show about one-third as mature and old growth in excess of 100 years (Rowe 1983, 1993).

Conclusion

Within the workshop that formed the core of this book (see Preface), several issues were raised that relate directly to the landscape level of organization (see also Chapter 1). This chapter has tried to address four of these issues directly. Summary comments on the issues follow.

- **Issue:** Management at the landscape level is hampered by lack of agreement on the definition of a landscape unit. Such units are difficult to define because landscapes are shifting mosaics of heterogeneous landforms, vegetation types, and land uses that are derived from both inherent environmental heterogeneity and natural and human-induced disturbances. Nonetheless, a broadly agreed definition must be reached before landscape diversity can be successfully incorporated into policy and management.
  Comment: The physiographic base of the land is the relatively stable stage on which the biotic actors come and go in their shifting scenes. As soil survey and forest site surveys show, the land can be divided hierarchically into meaningful units. Bounding such units and mapping their distributions is not arbitrary when guided by the purpose and knowledge of ecological relationships, especially those between vegetation and physiography (Host and Pregitzer 1992).

---

1   This is the approach taken in the *Biodiversity Guidebook* of the British Columbia Forest Practices Code (BC Forest Service and BC Environment 1995).

- **Issue:** Other issues at the regional landscape level needing resolution include lack of coordination among different resource interests and land users.
  Comment: The source of coordination and cooperation is the shared vision of a comprehensive and distant goal. "Sustaining biodiversity" is not enough. Maintaining a specified ecosystem mosaic structure and function, region by region, is better because it goes straight to the heart of ecological land-use planning. Moreover, it is measurable and therefore operational.

- **Issue:** Incompatibility exists between administrative and ecological units.
  Comment: Redraw the administrative unit boundaries or, "where natural landscape units have been administratively fragmented, management agencies and licensees must develop a strategic biodiversity plan together" (BC Forest Service and BC Environment 1995).

- **Issue:** There is disagreement on methods of maintaining mosaics (should landscape mosaics be maintained by natural disturbances such as pathogens or fire?).
  Comment: This is a question of purpose and of scale. If the goal is to maintain nature's creativity over large areas, then as part of that purpose natural disturbances and processes will proceed cyclically and be allowed to run their courses. In areas open to resource use, exploitation practices that mimic natural disturbances and processes would seem to have the best chance of proving sustainable over the long term. A mindset that is willing to learn from the natural world will also complement such exploitation practices, encouraging humanity to be less human-centred and more ecocentric.

The old Cartesian dualism positioned humans as the conscious thinking centre pitted against a dead and menacing world. In revealing our relationships to all other organic things, Darwin and the biologists that came after him softened this view but did not fundamentally change it. Organisms still maintain centre stage: biotic confronts abiotic, the animate confronts the relatively worthless environment.

The message is wrong. Life is a property of the whole – of the ecosphere and its ecosystem sectors, of the three-dimensional landscapes and waterscapes with all their matrix components – and not just of organisms alone. Earth spaces are endangered, and because of their evolutionary creativity they are more important over the long run than the species abstracted from them.

Without this insight conservationists keep slipping back into the language of saving organisms and their habitats, leaving the door open to those

who argue that the solution is more and better arks. To plan the preservation of biodiversity at the land or landscape level is to preserve more than
that term initially suggested. It is to preserve the world.

Literature Cited

Achuff, P.L., W.D. Holland, G.M. Coen, and K. Van Tighem. 1984. Ecological land classification of Mount Revelstoke and Glacier National Parks, British Columbia. Vol. 1: Integrated resource description. Alberta Institute of Pedology Publication No. M-84-11,
Edmonton, AB.

Albert, D.A., S.R. Denton, and B.V. Barnes. 1986. Regional landscape ecosystems of Michigan. School of Natural Resources and Environment, University of Michigan, Ann Arbor,
MI.

BC Forest Service and BC Environment. 1995. Biodiversity guidebook: Forest practices code
of British Columbia. Ministry of Forests, Victoria, BC.

Barnes, B.V. 1993. The landscape ecosystem approach and conservation of endangered
spaces. Endangered Species Update 10(3&4):13-19.

Blouin, G. 1993. The other 88 percent. Forestry Chronicle 69(2):114.

Boyle, T.J.B. 1992. Biodiversity of Canadian forests: Current status and future challenges.
Forestry Chronicle 68:444-53.

Burton, P.J., A.C. Balisky, L.P. Coward, S.G. Cumming, and D.D. Kneeshaw. 1992. The value
of managing for biodiversity. Forestry Chronicle 68:225-37.

Canadian Council of Forest Ministers. 1992. Sustainable forests, a Canadian commitment.
National Forest Strategy, Canadian Council of Forest Ministers, Hull, PQ.

Canadian Pulp and Paper Association. 1993. Old growth forests. Pp. 37-39 *in* G. Blouin and
R. Comeau (eds.). Forestry on the hill, Special Issue No. 5, Old Growth Forests. Canadian
Forestry Association, Ottawa, ON.

Demarchi, D.A., R.D. Marsh, A.P. Harcombe, and E.C. Lea. 1990. Pp. 55-144 *in* R.W. Campbell,
N.K. Dawe, I. McTaggart-Cowan, J.M. Cooper, G.W. Kaiser, and M.C.E. McNall (eds.). The
birds of British Columbia, Vol. 1. Royal British Columbia Museum, Victoria, BC.

Drury, W.H. 1969. Discussion on concepts of ecosystem and landscapes (an excerpt). P. 78
in K.N.H. Greenidge (ed.). Essays in plant geography and ecology. Nova Scotia Museum,
Halifax, NS.

Hammond, H. 1991. Seeing the forest among the trees: The case for wholistic forest use.
Polestar Publishers, Vancouver, BC.

Host, G.E., and K.S. Pregitzer. 1992. Geomorphic influences on ground flora and overstory
composition in upland forests of northwestern lower Michigan. Canadian Journal of
Forest Research 22:1547-55.

Ironside, G.R. 1991. Ecological land survey: Background and general approach. Pp. 3-9 *in*
H.A. Stelfox, G.R. Ironside, and J. Kansas (eds.). Guidelines for the integration of wildlife
and habitat evaluations with ecological land survey. Wildlife Working Group, Canada
Committee on Ecological Land Classification and Wildlife Habitat, Ottawa, ON.

Jenny, H. 1941. Factors of soil formation. McGraw-Hill, New York, NY.

--. 1980. The soil resources: Origin and behaviour. Springer-Verlag, New York, NY.

Kavanagh, K., and T. Iacobelli. 1995. A protected areas gap analysis methodology: Planning
for the conservation of biodiversity. World Wildlife Fund Canada, Toronto, ON.

Klopatek, C.C., E.G. O'Neill, D.W. Freckman, C.S. Bledsoe, D.C. Coleman, D.A. Crossley,
Jr., E.R. Ingham, D. Parkinson, and J.M. Klopatek. 1992. The sustainable biosphere initia-
tive: A commentary from the US Soil Ecology Society. Bulletin of the Ecological Society of
America 73:223-28.

MacKinnon, A., D. Meidinger, and K. Klinka. 1992. Use of the biogeoclimatic ecosystem
classification system in British Columbia. Forestry Chronicle 68:100-20.

Meidinger, D., and J. Pojar. 1991. Ecosystems of British Columbia. Research Branch, BC
Ministry of Forests, Victoria, BC.

Mosquin, T., P.G. Whiting, and D.E. McAllister. 1994. Canada's biodiversity: The variety of life, its status, economic benefits, conservation costs and unmet needs. The Canadian Centre for Biodiversity, Canada Museum of Nature, Ottawa, ON.

Naess, A. 1989. Ecology, community and lifestyle: Outline of an ecosophy. Cambridge University Press, New York, NY.

Noss, R.F. 1990. Can we maintain biological and ecological integrity? Conservation Biology 4:241-43.

–. 1992a. The wildlands project. Wild Earth, Special Issue:10-25.

–. 1992b. A preliminary biodiversity conservation plan for the Oregon Coast Range. Report to the Coast Range Association, Newport, OR.

Odum, E.P. 1970. Optimum population and environment: A Georgia microcosm. Current History 58:355-59.

Odum, E.P., and H.T. Odum. 1972. Natural areas as necessary components of man's total environment. Proceedings of the North American Wildlife and Natural Resources Conference, 12-15 March 1972, Mexico City, 37:178-89.

Rowe, J.S. 1983. Concepts of fire effects on plant individuals and species. Pp. 135-54 *in* R.A. Wein and D.A. MacLean (eds.). The role of fire in northern circumpolar ecosystems. SCOPE Vol. 18, John Wiley and Sons, Toronto, ON.

–. 1992a. Biological fallacy: Life equals organisms. BioScience 42(6):394.

–. 1992b. The integration of ecological studies. Functional Ecology 6:115-19.

Rowe, J.S., and B.V. Barnes. 1994. Bio-ecosystems and geo-ecosystems. Bulletin of the Ecological Society of America 75(1):40-41.

Rowe, S. 1993. Old growth forests. Pp. 26-28 *in* G. Blouin and R. Comeau (eds.). Forestry on the hill, Special Issue No. 5, Old Growth Forests. Canadian Forestry Association, Ottawa, ON.

Scheiner, S.M., A.J. Hudson, and M.A. VanderMeulen. 1993. An epistemology for ecology. Bulletin of the Ecological Society of America 74:17-21.

Seymour, R.S., and M.L. Hunter, Jr. 1992. New forestry in eastern spruce-fir forests: Principles and applications to Maine. Maine Agricultural Experiment Station Miscellaneous Publication 716. Bangor, ME.

Sims, R.A. 1992. Forest site classification in Canada: A current perspective. Forestry Chronicle 68:21-120.

Solbrig, O.T. 1991. The origin and function of biodiversity. Environment 33:16-20, 34-38.

Troll, C. 1971. Landscape ecology (geo-ecology) and bio-coenology – a terminology study. Geoforum 8:43-46.

7
At What Scale Should We Manage Biodiversity?
Reed F. Noss

Introduction

At a biodiversity workshop sponsored by the United States Bureau of Land Management (BLM) in Redding, California, I received an alarmingly negative response to my lectures on biodiversity and scales of management concern. The workshop participants included not only employees of the federal and state agencies that have land management authority in northern California but also members of local "bioregional councils." I have always had high hopes for the bioregional approach as championed, for example, by poet Gary Snyder. How can we take on the problems of the world if we can't even dwell peacefully, non-destructively, and self-sufficiently in our own bioregions? We need to know, love, and defend our home turf. Organized around watersheds, mountain ranges, vegetation types, and other natural boundaries, the bioregion is a sensible scale for biodiversity management – certainly more sensible than the conventional site level and also more "real" than regions defined by county lines or other political boundaries. The members of the bioregional councils at this workshop were a curious mix of environmentalists, timber workers, county commissioners, and other local people.

My message in those lectures in Redding, which I repeat here, was that biodiversity presents an opportunity to think big in space, time, level of organization, and ambition. Instead of managing land simply at the scale of the individual site or timber stand, we can step back and view each site in the context of the broader landscape in which it functions as an interacting component. Rather than planning at bureaucratic 5- or 10-year intervals – or, perhaps more commonly, until the next election – we can plan within the more meaningful scales of ecological and evolutionary time. Instead of focusing only on the species level of organization, we can consider a hierarchy of genetic, population, species, community, ecosystem, and landscape levels. And we can think big in terms of ambition, trying to accomplish more with conservation than we ever thought possible.

This "think big" message goes over well with most audiences, so I was unprepared for the angry responses I received from the bioregionalists. "We don't care about what's going on in Florida, the Rocky Mountains, or southern California," one woman said. "We have serious problems we need to deal with right here in our own watershed." A man chimed in, "Why do we need to hear about mass extinctions in the geological past, human overpopulation, or global warming? We have to put together a consensus response to the draft forest plan!" Apparently I was not a very effective communicator that day, for these people missed the whole point of providing context for management. I tried to explain that experiences elsewhere have something to teach us and that broader environmental problems such as human overpopulation and atmospheric changes will likely affect each of our little watersheds. But they would hear nothing of it. Although preaching "think globally, act locally," these bioregionalists were in fact thinking and acting very locally and narrowly.

My experience in Redding confirmed my suspicion that many people are not ready for the "big picture" world view of modern conservation, nor are they easily able to appreciate issues of scale or context. This situation has to change if we are ever to succeed in our mission of protecting and restoring biodiversity locally and globally.

My purpose here is to explore the question "At what scale should we study and manage biodiversity to maintain its ever-changing complexity?" My answer is that no single scale of analysis or planning will suffice for all our management needs. Different problems require resolution at different scales. We cannot reasonably consider long-term viability of grizzly bears within a 10,000 ha watershed, nor would we map the distribution of a narrowly endemic plant at a scale of 1:1,000,000. The trick is to match the management problem or question to the appropriate scale. Multi-dimensional problems – what we encounter, for example, when we attempt to devise a multi-species, multi-resource management plan for a broad region – require analysis and planning at many scales. "Think at multiple scales, act at multiple scales" would be my preferred motto for real-world conservation. Our conservation strategy should be hierarchical, with small-scale efforts embedded and interpreted within a broader context. For land managers, a landscape scale of thousands to tens of thousands of hectares is an efficient spatial scale for integrating concerns from many scales and evaluating progress in conservation.

Scales of Biodiversity Patterns and Processes

Let us first explore some scales at which patterns of biodiversity can be recognized and at which key ecological and evolutionary processes operate. Only then can we approach the question of what kind of management might be needed to perpetuate these patterns and processes. Conservationists today are interested in both pattern and process. Traditionally we were mostly concerned with the pattern of nature. For instance, we were interested in the scenic quality of natural landscapes, in spectacular natural features (such as geysers, canyons, and waterfalls), and in the recreational and tourism potential that such features possess.

Later, as science began to exert greater influence on conservation decision making, we became interested in the distribution of rare species, exemplary natural communities, and other elements of natural diversity. In the United States, The Nature Conservancy took the lead in mapping and evaluating these patterns (Jenkins 1985, 1988; Noss 1987), and the US Department of Interior is conducting a gap analysis of vegetation types and associated species in protected areas (Scott et al. 1993). In Canada, the World Wildlife Fund and the Canadian Council on Ecological Areas are assessing representation of "enduring features" of landscapes, based largely on topography and substrate. The aim of all these studies has been to capture the patterns of biodiversity – mostly at the species, community, and ecosystem levels of organization – in a representative system of reserves.

Increasingly, ecologists are emphasizing the dynamic qualities of natural ecosystems. The old idea of drawing a line around a site of high conservation value and trying to hold it forever in the condition in which we first found it has been rejected (Botkin 1990). Instead, we now talk about maintaining the geomorphological, ecological, and evolutionary processes that generate biodiversity and keep it forever changing over time. These processes include landform development, hydrological cycles, species migrations, isolation, speciation, gene flow, natural disturbances, succession, herbivory, predation, pollination, and many others. All of these processes operate at multiple scales. And of course, simply keeping these processes operating in some fashion is not so difficult – we could not stop them if we tried. Rather, conservationists are interested in maintaining processes within certain limits or ranges of variation that we call "natural," "historic," "acceptable," or "desirable." For example, if the species in a natural community have evolved in the presence of low-intensity fires occurring every 2 to 10 years, it makes sense to continue or simulate that intensity and frequency in a management program.

The reason we are interested in maintaining an acceptable range of variation in processes is simple: we have learned that when natural limits are exceeded, biodiversity is lost. That is, when processes escape their normal bounds, the patterns that we value are also disrupted. When fires are too rare or too frequent, too intense or not intense enough, many native species cannot cope with the change, and the natural community is altered. We can say the same thing about flooding, grazing, predation, gene flow, and many more. When natural processes are altered, species often become locally extinct; if the species are narrow endemics, they may become globally extinct. Extinction is as natural as speciation, but extinction rates hundreds or thousands of times greater than speciation have occurred only a few times in the geologic past (the mass extinction events recorded in the fossil record) and are not something conservationists want to see repeated today. Importantly, all of this is a matter of scale. Rate, intensity, and magnitude of processes are fundamental to the perpetuation of biodiversity.

At what spatial and temporal scales should we analyze the patterns and processes of biodiversity? If this were purely a question for basic scientific research, then the answer would be "at all scales at which these patterns and processes occur, from molecules to the entire biosphere." But we need not be so demanding or comprehensive when our mission is an applied one. Here we must know the scales at which the processes and patterns critical to the perpetuation of biodiversity can be practically measured, mapped, managed, and protected. Because we cannot measure everything, we rely on indicators whose status tells us about something bigger. We can consider the question of suitable spatial and temporal scales by examining the four levels of organization discussed in an earlier paper of mine on biodiversity indicators (Noss 1990).

Genetic Level

One might think that as we proceed up the biological hierarchy from genes to landscapes, our spatial scale of analysis also increases. This is not necessarily the case, however. Several scales of analysis and management are needed for each of our recognized levels of biological organization. Genetic analyses are conducted from the very small scale of molecules, as when we map genes along the DNA molecule, to global analyses of the distribution of genetic diversity within and among species. Genetic processes such as mutation that occur within chromosomes are as important as the global changes in selective pressures that ensue when the world's climate warms or cools.

For land managers, the genetic questions of greatest interest are those related to the genetic "structure" of species and the genetic aspects of population viability. The spatial scale at which we analyze these qualities depends on the spatial distribution of the species, the type of breeding system, and the level of genetic interchange between populations. As noted by Barrett and Kohn (1991):

> The total genetic variation maintained within a species can be partitioned in a hierarchical manner, according to the way it is distributed among regions, populations, and individuals within populations. Four evolutionary forces – mutation, natural selection, migration, and random genetic drift – interacting with an organism's recombination system, account for the manner in which variation is distributed among levels in the hierarchy. The

relative importance of these factors differs among species and ecological groups.

Narrowly endemic plants often have little genetic variation. For example, an electrophoretic analysis of 28 individuals in four populations of Furbish's lousewort (*Pedicularis furbishiae*), endemic to a small area in northern Maine and adjacent New Brunswick, detected no genetic variation whatsoever (Waller et al. 1987). Similarly, no allozyme variation or heterozygosity was observed in individuals from six populations of the rare *Bensoniella oregana* (Soltis et al. 1992). Most trees that have been well studied genetically seem to have high levels of genetic diversity, with most of the variation within single populations and with few differences among populations (Ledig 1986). Species with limited gene flow are expected to have greater differences among populations. Selfing species of plants generally have several times more genetic differentiation among populations than windpollinated outcrossing species (Hamrick et al. 1991). Some 96% of the genetic variation in the Butte County meadowfoam (*Limnanthes floccosa* subsp. *californica*) is distributed among populations; individuals within populations are essentially monomorphic, probably due to the high rate of selfing and a history of population bottlenecks (Dole and Sun 1992).

Most species of birds, because they can fly, have (like most trees) very few genetic differences among populations. On average, only 4.8% of the genetic variation of 23 bird species examined occurred among populations (Evans 1987). In contrast, the endangered Red-cockaded Woodpecker (*Picoides borealis*), apparently because of its cooperative breeding social structure, nonmigratory habits, and poor dispersal abilities, has roughly three times more differentiation among populations – about 14% of the detectable variation (Stangel et al. 1992). We can expect disjunct populations of plants and animals to have a high probability of being genetically distinct; some of them may be on the verge of becoming new species.

Thus the spatial scale for genetic management is highly species-specific. The temporal scale is also species-specific in that organisms with short generation times may change much more rapidly (from our perspective) in their genetic constitution than organisms with long generation times. We need to know something about the geographic distribution, dispersal capacities, breeding system, generation time, and population history of a species to manage it at the right spatial and temporal scale.

Usually a manager will not need to worry about genetic management except for species of commercial importance or those that are extremely rare. For many rare species, demographic factors, considered below, are a greater immediate threat to population viability than genetic problems (Lande 1988). Genetic considerations, however, do become crucial when planning over long time intervals – and we certainly need to plan over longer intervals! In many cases, decisions being made today about highway construction, urbanization, and other land development will affect the configuration and connectedness of habitats at a regional scale for decades and possibly centuries to come. Over this span of time, the effects of small population size and reduced gene flow on genetic integrity can be significant. Many small, isolated populations are expected to lose their evolutionary flexibility and their ability to adapt to a changing environment (Frankel and Soulé 1981).

The Florida panther (*Felis concolor coryi*) provides an example of how range declines can lead to genetic impoverishment. Government scientists responsible for conserving the Florida panther have focused their efforts on the tiny remnant population in south Florida while doing preliminary studies of potential reintroduction sites further north. Their concerns are immediate and their planning horizon relatively short. The precarious genetic status of the panther – reduced allozyme variation compared with other North American puma, over 90% defective sperm, cryptorchism, and other physical abnormalities (O'Brien et al. 1990; Fergus 1991) – is ultimately a consequence of two centuries or more of relentless persecution and habitat destruction throughout its original range, which spanned the southeastern United States. Furthermore, this subspecies of puma surely had genetic connections to the three other subspecies with which it was contiguous. The present inbred condition of the panther is almost certainly a result of small population size and absence of gene flow to and from other populations (Fergus 1991).

I emphasize that the trends that led to the genetic impoverishment and endangerment of the Florida panther were regional in scope. The reason the panther remains at all in south Florida is that until recently this was the last remote, roadless area within the panther's original range. The panther story provides a lesson for managing other wide-ranging species. The long-term viability of other large carnivores in North America – most notably the grizzly bear (*Ursus arctos horribilis*) – is being severely compromised today by the cumulative effects of logging, oil and gas development, and associated road building that fragments the original interlinked wilderness into small refugia surrounded by extensive mortality sinks (Shaffer 1992). The genetic integrity of grizzly bears in the American portion of their range may require a total population of some 2,000 animals (Allendorf et al. 1991) inhabiting some 13 million ha of interlinked wildlands (Metzgar and Bader 1992).

Population/Species Level

My comments on the population/species level largely mirror those for the genetic level. The spatial scale at which we manage a species must correspond to the scale at which the species is distributed and sometimes (in the case of rare animals) even the scale at which individuals use the environment.

Generally, we can manage plant or invertebrate populations at smaller scales than we must manage most vertebrates. The Nature Conservancy's original strategy of securing small, isolated reserves based on "element occurrences" (locations of rare species and natural communities) worked rather well because most of the elements mapped were rare plants. The heritage program (BC Conservation Data Centre) methodology of mapping point localities has been much more successful for rare plants, invertebrates, and small vertebrates than for species with larger area requirements. Among animals, area requirements usually increase with body size and trophic position, so that large carnivores tend to require the largest areas (Harestad and Bunnell 1979). But migratory ungulates such as bison or barren ground caribou may have even greater total area requirements. For both large carnivores and migratory ungulates, conservation planning and management must often encompass tens of millions of hectares (Noss 1992; Noss and Cooperrider 1994).

One might conclude that plants and small animals can be conveniently managed at the scale of the individual site – the management scale so familiar to us. But this may seldom be true. For one thing, many small animals (including rodents and many invertebrate groups) are characterized by extreme fluctuations in population size. High variance in population size is often correlated with susceptibility to extinction (Karr 1982; Pimm et al. 1988). C.D. Thomas (1990) noted that, for species with high levels of population variability, a population geometric mean must be 5,500 individuals to drop below 100 only once every 100 years. He suggested that a minimum viable population size of 1,000 should be adequate for species with normal levels of population fluctuation, and 10,000 should permit medium- to long-term persistence of birds and mammals with high levels of fluctuation. Invertebrate populations often fluctuate wildly. Herbivorous insects on *Phragmites australis* wetlands may require habitat patches big enough to support outbreak populations as large as 180,000 adults because of the extreme population changes characteristic of these insects (Tscharntke 1992).

Some of the best research on population persistence in small habitat patches has been done by Gray Merriam and students on Ontario farmland (see Chapter 4). In these landscapes, white-footed mouse (*Peromyscus leucopus*) populations in individual woodlots become extinct frequently. Vacated sites can be recolonized by dispersal of individuals from other woodlots, and dispersal occurs more readily when woodlots are connected by wooded fencerow corridors (Fahrig and Merriam 1985; Merriam 1988). Local extinctions of chipmunks (*Tamias striatus*) in these landscapes can also be reversed through colonization by individuals moving along fencerows from other woodlots (Henderson et al. 1985). Thus, even for small rodents, the scale at which population persistence must be considered is not the individual woodlot but rather a network of woodlots with various degrees of connectivity.

The Ontario farmland work offers excellent support for the metapopulation concept as a basis for management at the species level. I use the term *metapopulation* in a general sense to mean a system of populations connected by at least occasional dispersal, not necessarily implying any kind of equilibrium in the winking on and off of populations. Many so-called metapopulations lack balance between local extinction and colonization (Harrison 1991). Some species are distributed naturally as metapopulations, whereas in other cases formerly continuous populations have been converted to metapopulations by habitat fragmentation.[1]

The desirability of subdivided versus continuous populations has been debated since the SLOSS (single large or several small) controversy over reserve design (Soulé and Simberloff 1986). One advantage of having populations subdivided is that population extinctions may be independent of one another. A disease, for example, may wipe out one or two populations, but the metapopulation as a whole will persist if no infected individuals happen to disperse to the other populations. The disadvantages of subdivision include increased susceptibility to inbreeding depression, loss of alleles through genetic drift, demographic instability, and other problems that make smaller populations more vulnerable to extinction, all else being equal (Shaffer 1981; Quinn and Hastings 1987; Burkey 1989).

Some threshold level of dispersal among populations is necessary for metapopulations to persist, which is one of the many arguments for maintaining or restoring connectivity in landscapes (Merriam 1988, 1991; Noss 1993a; Chapter 4). If the rate of local extinctions is higher than the rate of patch recolonization through dispersal, the metapopulation will become extinct. Studies of the Bay checkerspot butterfly (*Euphydryas editha bayensis*) suggest that local extinction is frequent on small patches of serpentine grassland, to which the species is now restricted because of fragmentation of the original native grassland. Persistence of the metapopulation depends on dispersal to recolonize vacated patches. Because the Bay checkerspot is a relatively poor disperser, stepping-stone habitat patches that reduce isolation are important (Murphy and Weiss 1988). The metapopulation model also suggests that habitat patches currently unoccupied may be critical to survival, because they represent sites for possible recolonization. The spatial scale of a metapopulation depends heavily on the dispersal capabilities of individuals. If patches of suitable habitat are spaced further apart than the maximum dispersal distance of the species in question, then there is no metapopulation, just a constellation of isolated populations. If individuals are translocated among patches as part of a management strategy to prevent genetic isolation, we have an artificially maintained metapopulation.

---

1 G. Merriam provides a more complete discussion of metapopulations in Chapter 4 of this volume.

On the other hand, if patches of suitable habitat are so close together that individuals regularly travel between them, we have a single population rather than a metapopulation (Harrison 1991). This is obviously the case when the home ranges of individuals encompass several separate habitat patches, as is true for many birds and mammals.

In all these circumstances, the scale at which we manage the population or species must correspond to the scale at which the organisms obtain resources, disperse from the place they were born, find mates, and otherwise use the environment. Hence the need for good empirical research. Unfortunately, many metapopulation models are highly abstract and spatially inexplicit because the necessary data on dispersal and other aspects of autecology are absent. Usually, in the United States, data are available only for species that have been studied intensively because they are legally protected and their protection interferes in some way with economic activities. Studies of the threatened Northern Spotted Owl (*Strix occidentalis caurina*) determined that juveniles dispersed from 3 to 64 km from their natal territories; thus recommendations were made to space habitat conservation areas no further apart than the mean dispersal distance of about 32 km (Thomas et al. 1990). Research on the California Gnatcatcher (*Polioptila californica*), another threatened species, has documented natal dispersal from 0.45 to 9.8 km, with a mean distance of 2.8 km and a median of 2.2 km (Ogden Environmental and Energy Services 1992; P.J. Mock, personal communication, 1993). Not enough is known, however, about many other aspects of the autecology of the Gnatcatcher to develop a scientifically and legally defensible population viability analysis and conservation plan (California Department of Fish and Game 1993). For most other species, we know much less.

Community/Ecosystem and Landscape Levels
At what scale should we analyze and manage ecosystems? Ecologists are wont to point out that the boundaries of ecosystems are arbitrary and that ecosystems can be recognized at many spatial scales. A pool of water in a hollow tree bole is an ecosystem, as is the forest in which the tree lives, the biome in which the forest exists, and finally the entire biosphere. For management purposes, we often define ecosystem boundaries around vegetation types or habitats that can be remotely sensed, mapped, and manipulated in a geographic information system (GIS). I lump the community/ecosystem and landscape levels here because the landscape is essentially a heterogeneous ecosystem. Many conservationists use the term "greater ecosystem" to refer to large regional landscapes of conservation interest, such as the 7.7 million ha Greater Yellowstone Ecosystem (Goldstein 1992).

Species richness and diversity are essentially community-level concepts, although they can be measured at spatial scales from microscopic (for bacteria) to global. While species richness increases continuously as we increase the area sampled (Connor and McCoy 1979), this does not mean it increases steadily; a species-area curve can have some humps in it. For example, if we sampled vascular plants or birds in southwestern British Columbia, we would approach a plateau in species richness as more and more area is sampled. But if we expanded our sampling eastward over the crest of the Cascade Mountains, we would encounter a burst in species richness as we suddenly found species that were adapted to the drier "eastside" forests. We could predict similar stepwise increases in species richness as we progressed up and over the Rocky Mountains. Species turnover is higher when environmental gradients are steeper.

In species-area studies it is helpful to note the spatial scale of observation and how species composition changes from one scale to another. The collection of species within an area of relatively uniform habitat is called *alpha diversity* or *within-habitat diversity* (Whittaker 1972; Karr 1976). Physically similar habitats in the same region can be expected to have similar species composition and alpha diversity. As we expand the area sampled, we move along environmental gradients (such as upslope, downslope, or from one soil type to another) and encounter new species adapted to these different conditions. The turnover in species along an environmental gradient is called *beta diversity* or *between-habitat diversity*. At a still broader scale, many environmental gradients and geographic replacements of species occur as range boundaries are crossed. Diversity at this regional scale is called *gamma diversity*. The alpha, beta, and gamma diversity concepts are useful for comparing biodiversity in different regions or in the same region under different management scenarios (Noss and Cooperrider 1994). Importantly, increases in species richness at a site or landscape scale caused by habitat fragmentation (artificial beta diversity) may be accompanied by a reduction of species richness at a regional scale, as species sensitive to human activities are progressively lost and each landscape becomes dominated by opportunistic weeds (Noss 1983).

Species richness, diversity, and composition do not change smoothly as we change spatial scale because virtually all landscapes are mosaics at one scale or another. At a landscape scale of analysis (a few kilometres across), the distribution of vegetation types typically corresponds to changes in elevation and topographic position. A "gradient mosaic" of vegetation represents responses of species to these environmental gradients and is especially obvious in mountainous regions, as shown in Figure 7.1 (Whittaker 1956; Peet 1988).

On top of the environmentally determined gradients and mosaics of vegetation are other mosaics created by natural disturbances and subsequent succession. The "grain" of a landscape is determined largely by the spatial scale of disturbance, that is, by the size and distribution of disturbance-generated patches. Relatively large disturbances, such as extensive fires,

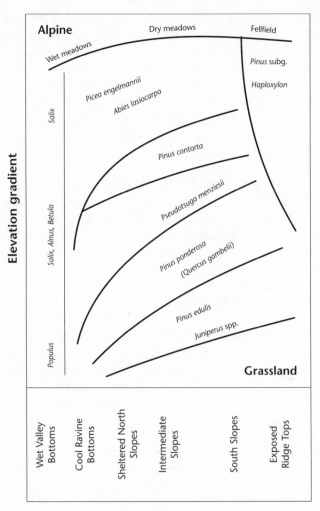

Topographic-moisture gradient

Figure 7.1   A gradient mosaic of vegetation types
responding to elevation and topographic-moisture
gradients in the central Rocky Mountains (from Peet
1988).

create a coarse-grained pattern, whereas canopy gaps caused by death and
fall of individual trees or small groups of trees create fine-grained patterns.
Because any landscape will be affected by many different kinds and scales
of disturbance, several grains of pattern may be overlaid on one another.
And, of course, what is a coarse-grained pattern for a slug may be a fine-
grained pattern for a moose.

Disturbances are typically patchy in time and space, so that new distur-bances occur in some areas while previously disturbed sites are recovering in others. This continuously changing pattern has been called a *space-time mosaic* (Watt 1947) or *shifting mosaic* (Bormann and Likens 1979). The spa-tial and temporal scales of this shifting mosaic have implications for reserve design and management. Reserves that are small relative to the spatial scale (patch size) of disturbance may experience dramatic fluctuations in the pro-portions of different seral stages over time, which in turn may threaten populations that depend on certain stages.

If a reserve is to maintain a reasonably stable mix of seral stages and spe-cies, it must be sufficiently large that only a relatively small part of it is disturbed at any one time. A source of colonists must also exist within the reserve or within dispersal distance so that populations can be re-established on disturbed sites. Pickett and Thompson (1978) defined a "minimum dy-namic area" as "the smallest area with a natural disturbance regime, which maintains internal recolonization sources, and hence minimizes extinction." In theory, a minimum dynamic area should be able to manage itself, main-taining habitat diversity and associated native species with no human in-tervention. Although few, if any, reserves in vegetation types subject to large, stand-replacing disturbances encompass minimum dynamic areas, biodiversity in reserves should be more secure when the disturbance inter-val is long compared with recovery time and when only a small portion of the landscape is affected. While no reserve size can guarantee stability, larger reserves have a lower probability of major shifts in landscape dynamics caused by rare disturbance events (Turner et al. 1993).

In this discussion I have emphasized biological responses to environmen-tal gradients and disturbance regimes because they are of such paramount importance to biodiversity. But other abiotic and biotic processes – hydro-logic regimes, nutrient cycles, herbivory, predation, pollination, seed dis-persal, and so on – have characteristic spatial and temporal scales of operation in each regional ecosystem. A comprehensive program for conserving biodiversity would seek to determine these relevant scales.

Implementation

Biodiversity is complex, but from a management standpoint not hopelessly complex. Using the basic components and tools already within our grasp, we have a good chance of success with an integrated conservation strategy for biodiversity. But we have to think big. We must be able and willing to zoom back and forth from the microscope to the macroscope, from mites in the leaf litter to biomes on the continents.

Perhaps moving at ease from the leaf litter to the biosphere is too much to ask of anyone. But while I believe we should try to conceptualize and comprehend nature at all the scales at which natural phenomena operate, a practical conservation strategy can be built on the spatial scales at which the most important processes and patterns of biodiversity can be measured and mapped. For example, I have been involved in a conservation planning effort in southern California focusing on the coastal sage scrub and its many rare species, the most influential of which is the federally listed California Gnatcatcher mentioned earlier. Because not enough biological data are presently available to develop a scientifically or legally defensible conservation plan, our Scientific Review Panel presented interim conservation guidelines (California Department of Fish and Game 1993) to cover the three- to six-year period during which data sufficient to construct a more permanent plan will be collected.

Both the biological picture and the planning issues in southern California are highly complex. The Scientific Review Panel was able, however, to identify three spatial scales at which research could be conducted and data organized to answer the most pressing questions. The broadest scale is the entire planning region of southwestern California below 1,067 m elevation (above which coastal sage scrub does not occur). Over this area, the vegetation, vertebrates, and rare plants and animals have been mapped by the California Gap Analysis and California Natural Diversity Data Base at scales of 1:100,000 and 1:250,000. At this scale, with a minimum mapping unit (MMU) of 100 ha and a minimum resolution of 100 m, the following research questions can be addressed:

- What is the general distribution of coastal sage scrub in relation to other habitat types within the region?
- How does the present distribution of coastal sage scrub correlate with physical geographic variables such as physiography, elevation, topography, bedrock geology, soils, distance from coast, and watersheds?
- How does the present distribution of coastal sage scrub compare with the natural or historical distribution?
- How might the present distribution of coastal sage scrub be partitioned, for practical planning purposes, into subregions?
- Where are the large, relatively contiguous blocks or clusters of coastal sage scrub that might serve as core areas in a regional reserve network?
- What is the distribution of rare species and groups of species across the region and in adjacent regions?
- What is the protection status of large blocks of coastal sage scrub in the region?
- Which landscapes might serve as linkages between large blocks of coastal sage scrub?

Several of these questions have been addressed by the Scientific Review Panel, the US Fish and Wildlife Service, the California Department of Fish and

Game, and the California Gap Analysis Project. I suggest these questions are applicable to any planning region.

The intermediate scale of analysis we recommended includes 13 planning subregions identified on the basis of an earlier regional analysis of clusters of coastal sage scrub (Figure 7.2). At this planning scale, the 1:24,000 scale (7.5-minute) topographic maps produced by the US Geological Survey are useful and an MMU of 2-4 ha is reasonable for most vegetation types. For subregions, the questions listed above can be addressed in much greater detail over a smaller area. For instance, populations of rare species can often be mapped as polygons rather than as points. Small patchy habitats (such as vernal pools) can actually be mapped rather than included as attributes in labels for larger vegetation polygons. The relationship of coastal sage scrub subassociations to elevation, slope aspect, and other physical habitat variables can be analyzed in detail. We provided guidelines for biological surveys of coastal sage scrub (Noss et al. 1992a) and solicited research proposals to answer remaining questions. Unfortunately, funding was never provided for the biological surveys we called for, and many critical research questions remain unanswered (Noss et al. 1997). Meanwhile development continues.

Finally, the finest scale for research and management is that of individual sites. When decisions are being made that will affect what a landowner can or cannot do on a piece of property, map-based information has to be precise and reliable enough to stand up in court. For this purpose, a 1:4,800 map scale is suitable. At this scale, vegetation patterns, locations of individual rare plants and animal territories, and other biological and physical data are mapped on the basis of detailed ground surveys and aerial photographs.

These three planning scales (or something roughly equivalent, depending on availability of maps) are likely to be applicable to many conservation and regional planning efforts. At each scale, indicators of biodiversity that can be measured and analyzed to tell us something about the broader ecosystem should be selected. By monitoring these indicators, we can gain at least some impression of how biodiversity as a whole might be changing over time and responding to alternative management treatments. This feedback allows us to learn from our management experiences and adjust our practices accordingly – the so-called adaptive management strategy (Holling 1978; Walters 1986).

Elsewhere I have presented a general framework for monitoring biodiversity at several levels of organization (Noss 1990). Briefly, the common assumption that one or a few indicator species can represent the needs of all species associated with the same habitat is fallacious (Landres et al. 1988). For example, the California Gnatcatcher is an incomplete umbrella species for the biodiversity of the coastal sage scrub; thus, research and

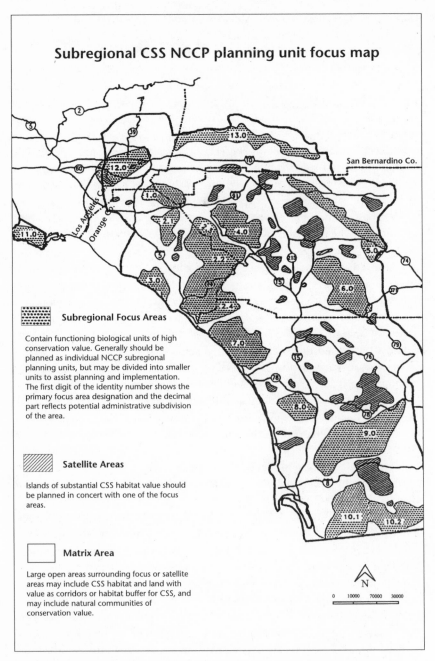

Subregional CSS NCCP planning unit focus map

San Bernardino Co.

Subregional Focus Areas

Contain functioning biological units of high
conservation value. Generally should be
planned as individual NCCP subregional
planning units, but may be divided into smaller
units to assist planning and implementation.
The first digit of the identity number shows the
primary focus area designation and the decimal
part reflects potential administrative subdivision
of the area.

Satellite Areas

Islands of substantial CSS habitat value should
be planned in concert with one of the focus
areas.

Matrix Area

Large open areas surrounding focus or satellite
areas may include CSS habitat and land with
value as corridors or habitat buffer for CSS, and
may include natural communities of
conservation value.

N

0   10000   70000   30000

Figure 7.2 Subregions recognized for the coastal sage scrub (CSS) natural
community conservation planning (NCCP) process in southern California (from
California Department of Fish and Game, unpublished).

monitoring guidelines for the coastal sage scrub focus on a suite of indicators (Noss et al. 1992a). Whenever possible we need to monitor demographic parameters of species (with special attention to keystone, umbrella, and sensitive species) as well as characteristics of their habitats. More broadly, it is a good idea to monitor structural, functional, and compositional indicators at each level of organization (Noss 1990).

Integration at a Landscape Scale
I have emphasized that the scale of conservation planning and management must match the scales at which critical ecological phenomena operate. I have described three spatial scales for planning and research, applied in a case study in California, that should address most phenomena of interest. For the land manager, however, I suggest that planning, managing, and monitoring multiple biological phenomena can be conveniently integrated at a "landscape" spatial scale of thousands to hundreds of thousands of hectares. The boundaries of the landscape should ideally match natural features such as watersheds or mountain ranges. This spatial scale provides the context for considering specific issues, such as designing forestry operations that do not threaten the population viability of given species. The landscape, because it is nested within a series of ever larger natural regions, is also a bridge to broader spatial scales where other phenomena operate and over which the population viability of the widest-ranging species, such as grizzly bears, must be considered.

At this landscape scale, we can follow a four-step process of conservation management. The first step is to describe the present state of the landscape and determine how it differs from the natural or historical condition, as shown in Figure 7.3 (Noss et al. 1992b; Noss 1993b, 1993c). Specifically, what trends or changes in this landscape have been associated with loss of biodiversity and ecological integrity? The seven trends shown in Figure 7.3 for a hypothetical forest landscape represent a shift in landscape structure to a present condition that is less favourable to native biodiversity than the natural condition. Although natural condition cannot be easily defined and can change over time, we do not have to quantify natural condition to understand how past and present trends associated with human activity threaten biodiversity. Compared with a natural landscape, the present condition of many managed forest landscapes is one of younger forests, reduced structural diversity within stands, smaller and more isolated patches of forest, fewer natural fires, and an increased density of roads. These trends are well documented in many forested regions and have been largely responsible for endangering species (the bottom arrow or dependent variable in Figure 7.3) and for other symptoms of biotic impoverishment (Norse 1990; Noss 1993b, 1993c).

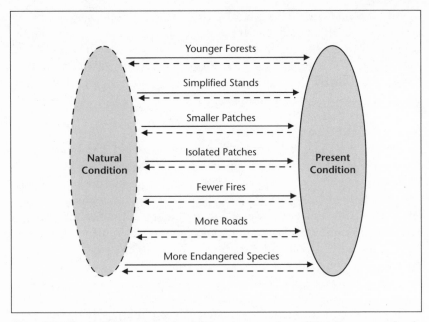

Figure 7.3   The shift from natural condition to present condition for a generalized forest landscape as exemplified by seven trends (Noss et al. 1992b; Noss 1993b, 1993c). Dashed arrows in reverse direction to trends indicate the potential for restoration.

The second step is to determine where we want to go, that is, to set goals and objectives. I disagree with the popular idea of describing a "desired future condition" for a managed landscape, at least in any specific sense. Nature is too variable and unpredictable for us to specify a future condition. Rather, I believe we should identify the desired direction of landscape change, monitor the response of the system to our progress in that direction, and decide later when the response has been adequate and the system has recovered. In other words, the goal is to restore the landscape to integrity by reversing the trends that have been associated with impoverishment. The goal of reversing degenerative trends does not imply returning to some pristine natural condition; such is neither possible nor desirable in most cases. Rather, we simply try to move in the direction of the dashed-line arrows in Figure 7.3, thereby removing or mitigating the factors that threaten biodiversity.

The third step is to specify the management actions that have a high probability of moving the landscape in the direction of recovery. Again referring to the dashed-line arrows in Figure 7.3, restoration may require the following:

(1) moving the landscape towards a more natural or equitable mix of seral stages by preserving existing late-successional stands, allowing many young stands to mature, and managing plantations on longer rotations

(2) retaining structural diversity, including snags and downed logs, in managed stands

(3) managing for large, intact patches of forest unfragmented by roads, clearcuts, or other openings

(4) retaining or restoring linkages between forests and providing connectivity across the regional landscape as a whole

(5) allowing natural fires to burn, using prescribed fire, or applying silvicultural manipulations that simulate fire and other disturbances so that we maintain a full spectrum of natural seral stages and structures

(6) stopping or dramatically reducing road construction and reconstruction, and obliterating and revegetating existing roads wherever possible

(7) recovering viable populations of rare species and reintroducing extirpated species.

Because we do not know precisely which management practices will move us fastest in the desired direction, it is essential that managers apply several experimental treatments in an adaptive management framework (Holling 1978; Walters 1986).

The fourth step is to determine how the landscape and its biodiversity respond to our management experiments and then modify our management accordingly. Thus we will need to revisit steps 1, 2, 3, and 4 in a continuous feedback cycle. We will need to monitor a broad suite of biodiversity indicators (Noss 1990) to obtain the information upon which changes in policies or management plans must be based. Because everything we do in land management and conservation generally is an experiment with an uncertain outcome (Walters and Holling 1990), I believe we should experiment very cautiously. Caution requires leaving plenty of unmanipulated wild areas as controls for our experiments and, whenever we are in doubt, being willing to risk erring on the side of protecting too much.

Literature Cited

Allendorf, F.W., R.B. Harris, and L.H. Metzgar. 1991. Estimation of effective population size of grizzly bears by computer simulation. Pp. 650-54 *in* E.C. Dudley (ed.). The unity of evolutionary biology: Proceedings of the Fourth International Congress of Systematic and Evolutionary Biology. Dioscorides Press, Portland, OR.

Barrett, S.C.H., and J.R. Kohn. 1991. Genetic and evolutionary consequences of small population size in plants: Implications for conservation. Pp. 3-30 *in* D.A. Falk and K.E. Holsinger (eds.). Genetics and conservation of rare plants. Oxford University Press, New York, NY.

Bormann, F.H., and G.E. Likens. 1979. Pattern and process in a forested ecosystem. Springer-Verlag, New York, NY.

Botkin, D.B. 1990. Discordant harmonies: A new ecology for the twenty-first century. Oxford University Press, New York, NY.

Burkey, T.V. 1989. Extinction in nature reserves: The effect of fragmentation and the importance of migration between reserve fragments. Oikos 55:75-81.

California Department of Fish and Game. 1993. Southern California coastal sage scrub natural community conservation planning. Draft conservation guidelines. 20 July 1993. Sacramento, CA.

Connor, E.F., and E.D. McCoy. 1979. The statistics and biology of the species-area relationship. American Naturalist 113:791-833.

Dole, J.A., and M. Sun. 1992. Field and genetic survey of the endangered Butte County meadowfoam – *Limnanthes floccosa* subsp. *californica* (Limnanthaceae). Conservation Biology 6:549-58.

Evans, P.G.H. 1987. Electrophoretic variability and gene products. Pp. 105-62 *in* F. Cooke and P.A. Buckley (eds.). Avian genetics: A population and ecological approach. Academic Press, Orlando, FL.

Fahrig, L., and G. Merriam. 1985. Habitat patch connectivity and population survival. Ecology 66:1762-68.

Fergus, C. 1991. The Florida panther verges on extinction. Science 251:1178-80.

Frankel, O.H., and M.E. Soulé. 1981. Conservation and evolution. Cambridge University Press, Cambridge, UK.

Goldstein, B.E. 1992. Can ecosystem management turn an administrative patchwork into a Greater Yellowstone Ecosystem? Northwest Environmental Journal 8:285-324.

Hamrick, J.L., M.J.W. Godt, D.A. Murawski, and M.D. Loveless. 1991. Correlations between species traits and allozyme diversity: Implications for conservation biology. Pp. 75-86 *in* D.A. Falk and K.E. Holsinger (eds.). Genetics and conservation of rare plants. Oxford University Press, New York, NY.

Harestad, A.S., and F.L. Bunnell. 1979. Home range and body weight – a re-evaluation. Ecology 60(2):389-402.

Harrison, S. 1991. Local extinction in a metapopulation context: An empirical evaluation. Biological Journal of the Linnean Society 42:73-88.

Henderson, M.T., G. Merriam, and J. Wegner. 1985. Patchy environments and species survival: Chipmunks in an agricultural mosaic. Biological Conservation 31:95-105.

Holling, C.S. (ed.). 1978. Adaptive environmental assessment and management. John Wiley and Sons, New York, NY.

Jenkins, R.E. 1985. Information methods: Why the heritage programs work. The Nature Conservancy News 35(6):21-23.

–. 1988. Information management for the conservation of biodiversity. Pp. 231-39 *in* E.O. Wilson (ed.). Biodiversity. National Academy Press, Washington, DC.

Karr, J.R. 1976. Within- and between-habitat avian diversity in African and neotropical lowland habitats. Ecological Monographs 46:457-81.

–. 1982. Population variability and extinction in the avifauna of a tropical land bridge island. Ecology 63:1975-78.

Lande, R. 1988. Genetics and demography of biological conservation. Science 241:1455-60.

Landres, P.B., J. Verner, and J.W. Thomas. 1988. Ecological uses of vertebrate indicator species: A critique. Conservation Biology 2:316-28.

Ledig, F.T. 1986. Heterozygosity, heterosis, and fitness in outbreeding plants. Pp. 77-104 *in* M.E. Soulé (ed.). Conservation biology: The science of scarcity and diversity. Sinauer, Sunderland, MA.

Merriam, G. 1988. Landscape dynamics in farmland. Trends in Ecology and Evolution 3:16-20.

–. 1991. Corridors and connectivity: Animal populations in heterogeneous environments. Pp. 133-42 *in* D.A. Saunders and R.J. Hobbs (eds.). Nature conservation 2: The role of corridors. Surrey Beatty and Sons, Chipping Norton, NSW, Australia.

Metzgar, L.H., and M. Bader. 1992. Large mammal predators in the northern Rockies: Grizzly bears and their habitat. Northwest Environmental Journal 8:231-33.

Murphy, D.D., and S.B. Weiss. 1988. Ecological studies and the conservation of the Bay checkerspot butterfly, *Euphydryas editha bayensis*. Biological Conservation 46:183-200.

Norse, E.A. 1990. Ancient forests of the Pacific Northwest. The Wilderness Society and Island Press, Washington, DC.

Noss, R.F. 1983. A regional landscape approach to maintain diversity. BioScience 33:700-6.

–. 1987. From plant communities to landscapes in conservation inventories: A look at The Nature Conservancy (USA). Biological Conservation 41:11-37.

–. 1990. Indicators for monitoring biodiversity: A hierarchical approach. Conservation Biology 4:355-64.

–. 1992. The Wildlands Project: Land conservation strategy. Wild Earth (Special Issue):10-25.

–. 1993a. Wildlife corridors. Pp. 43-68 *in* D.S. Smith and P.C. Hellmund (eds.). Ecology of greenways. University of Minnesota Press, Minneapolis, MN.

–. 1993b. A conservation plan for the Oregon Coast Range: Some preliminary suggestions. Natural Areas Journal 13:276-90.

–. 1993c. Sustainable forestry or sustainable forests? Pp. 17-43 *in* G.H. Aplet, N. Johnson, J.T. Olson, and V.A. Sample (eds.). Defining sustainable forestry. The Wilderness Society and Island Press, Washington, DC.

Noss, R.F., and A.Y. Cooperrider. 1994. Saving nature's legacy: Protecting and restoring biodiversity. Defenders of Wildlife and Island Press, Washington, DC.

Noss, R.F., J. O'Leary, D. Murphy, P. Brussard, and M. Gilpin. 1992a. Coastal sage scrub survey guidelines. Southern California Coastal Sage Scrub Scientific Review Panel. Stanford University, Stanford, CA.

Noss, R.F., S.P. Cline, B. Csuti, and J.M. Scott. 1992b. Monitoring and assessing biodiversity. Pp. 67-85 *in* E. Lykke (ed.). Achieving environmental goals: The concept and practice of environmental performance review. Belhaven Press, London, UK.

Noss, R.F., M.A. O'Connell, and D.D. Murphy. 1997. The science of conservation planning: Habitat conservation under the Endangered Species Act. Island Press, Washington, DC.

O'Brien, S.J., M.E. Roelke, N. Yuhki, K.W. Richards, W.E. Johnson, W.L. Franklin, A.E. Anderson, O.L. Bass, R.C. Belden, and J.S. Martenson. 1990. Genetic introgression within the Florida panther *Felis concolor coryi*. National Geographic Research 6:485-94.

Ogden Environmental and Energy Services Company, Inc. 1992. Task 3.5a-1. Accounts of MSCP target species. Unpublished draft report to Clean Water Program, City of San Diego, CA.

Peet, R.K. 1988. Forests of the Rocky Mountains. Pp. 63-101 *in* M.G. Barbour and W.D. Billings (eds.). North American terrestrial vegetation. Cambridge University Press, New York, NY.

Pickett, S.T.A., and J.N. Thompson. 1978. Patch dynamics and the design of nature reserves. Biological Conservation 13:27-37.

Pimm, S.L., H.L. Jones, and J. Diamond. 1988. On the risk of extinction. American Naturalist 132:757-85.

Quinn, J.F., and A. Hastings. 1987. Extinction in subdivided habitats. Conservation Biology 1:198-208.

Scott, J.M., F. Davis, B. Csuti, R. Noss, B. Butterfield, C. Groves, J. Anderson, S. Caicco, F. D'Erchia, T.C. Edwards, J. Ulliman, and R.G. Wright. 1993. Gap analysis: A geographical approach to protection of biological diversity. Wildlife Monographs 123:1-41.

Shaffer, M.L. 1981. Minimum population sizes for species conservation. BioScience 31:131-34.

–. 1992. Keeping the grizzly bear in the American West: A strategy for real recovery. The Wilderness Society, Washington, DC.

Soltis, P.S., D.E. Soltis, T.L. Tucker, and F.A. Lang. 1992. Allozyme variability is absent in the narrow endemic *Bensoniella oregana* (Saxifragaceae). Conservation Biology 6:131-34.

Soulé, M.E., and D. Simberloff. 1986. What do genetics and ecology tell us about the design of nature reserves? Biological Conservation 35:19-40.

Stangel, P.W., M.R. Lennartz, and M.H. Smith. 1992. Genetic variation and population structure of Red-cockaded Woodpeckers. Conservation Biology 6:283-92.

Thomas, C.D. 1990. What do real population dynamics tell us about minimum viable population sizes? Conservation Biology 4:324-27.

Thomas, J.W., E.D. Forsman, J.B. Lint, E.C. Meslow, B.R. Noon, and J. Verner. 1990. A conservation strategy for the Northern Spotted Owl. USDA Forest Service, USDI Bureau of Land Management, USDI Fish and Wildlife Service, and USDI National Park Service, Portland, OR.

Tscharntke, T. 1992. Fragmentation of *Phragmites* habitats, minimum viable population size, habitat suitability, and local extinction of moths, midges, flies, aphids, and birds. Conservation Biology 6:530-36.

Turner, M.G., W.H. Romme, R.H. Gardner, R.V. O'Neill, and T.K. Kratz. 1993. A revised concept of landscape equilibrium: Disturbance and stability on scaled landscapes. Landscape Ecology 8:213-27.

Waller, D.M., D.M. O'Malley, and S.C. Gawler. 1987. Genetic variation in the extreme endemic *Pedicularis furbishiae* (Scrophulariaceae). Conservation Biology 1:335-40.

Walters, C.J. 1986. Adaptive management of renewable resources. McGraw-Hill, New York, NY.

Walters, C.J., and C.S. Holling. 1990. Large-scale management experiments and learning by doing. Ecology 71:2060-68.

Watt, A.S. 1947. Pattern and process in the plant community. Journal of Ecology 35:12-22.

Whittaker, R.H. 1956. Vegetation of the Great Smoky Mountains. Ecological Monographs 26:1-80.

–. 1972. Evolution and measurement of species diversity. Taxon 21:213-51.

8
Setting Goals for Biodiversity in Managed Forests
Fred L. Bunnell

Before and during the workshop from which this book was derived (Preface), a list of 17 major issues related to monitoring biological diversity was generated; it is presented in Chapter 1 together with related background. This chapter summarizes steps that help address the issues. It relies on papers presented at the workshop (Chapters 1 through 7), discussions at the workshop, and other literature. As summarized here, actions to sustain biodiversity are general; specific management actions must acknowledge actual location and associated characteristics such as disturbance, land-use history, and forest type.

Issues in Measuring Biodiversity for Forest Policy and Management
The issues, phrased here as questions, are grouped under three broad headings: (1) Where are we going? (2) How do we get there? (3) How will we know when we get there? Actions to facilitate the monitoring of biodiversity involve policy, management, and research.

Where Are We Going?
This broad strategic question recognizes that activities of management and policy have a goal. In delineating that goal, three points are evident. First, existing definitions of biological diversity do not provide an operational target that translates neatly into forest attributes altered by managers (Bunnell 1994, 1998; Delong 1996; Chapter 1 of this volume). Second, components of the common definitions of biological diversity are themselves not unequivocally defined (Chapters 5 and 6). Third, the term *biological diversity* or *biodiversity* suffers from reification – the conversion of an abstract concept into a concrete thing or identifiable entity (Bunnell 1998; Chapter 4). Major issues influencing goal setting are interconnected; they follow as listed in Chapter 1.

(1) What Is the Goal?
This question is as much one of policy as of science. The scope of biological diversity yields a remarkably complex, largely unknown, and ill-defined goal. Biological diversity is currently ill defined as a scientific concept (Delong 1996; Bunnell 1998; Chapter 4); moreover, broad goals are inevitably set by policy rather than research (Chapter 2). Five steps can be taken: (a) separate definition and application of actions, (b) acknowledge societal goals, (c) link monitoring to goals, (d) focus research on key elements, and (e) eliminate from goals components that are difficult to measure or define. These are discussed below.

(a) *Separate definition and application of actions.* This step appears paradoxical but it is important. Delong (1996) reviewed 85 definitions of biological diversity and found that combining definition and application contributes to confusion about how to implement biodiversity concepts (see also Bunnell 1998). Much of this confusion derives from lack of agreement on what constitutes biological diversity – that is, a definition. Defining biological diversity is as elusive as defining ecological diversity, and argument about ecological diversity has continued since 1943. The application of management actions cannot await a tidy definition but should not attempt to limit the still poorly defined, fundamental meaning of biological diversity to more practically measurable goals. Although ill defined and poorly expressed in definitions, the concepts embraced by biological diversity are useful and powerful. Defining operational or management goals must be consistent with concepts embedded within the fundamental definition of biological diversity, but must proceed faster than will clarification of definitions and embedded concepts, and thus somewhat independently. Bunnell (1997a, 1998) argues that the appropriate approach for management and policy is to acknowledge those societal concerns that are consistent with the fundamental definition of biological diversity (see below). Interim management goals can focus on species, for example (questions 2 and 6), but must recognize that doing so takes into account only a limited meaning of biological diversity.

(b) *Acknowledge societal goals embraced by commitment to sustaining biological diversity instead of focusing on the elements of definitions.* Relate goals to underlying concerns. Delong (1996) and Bunnell (1998) found no agreement on the elements of definitions of biological diversity. The charge to sustain biological diversity, however, derives from genuine, thoughtful public concerns (described by Maini in Chapter 2). As Merriam noted in Chapter 4, the challenge for researchers or practitioners is not to devise definitions for ill-defined terms but to direct public and political energies to related issues of substance for which we have applicable knowledge. Bunnell (1998) summarized the public goals embedded in the Convention on Biological Diversity and related agreements emerging from the 1992 United Nations

Conference on Environment and Development (UNCED), or "Earth Summit," as:

- reduce rates of extinction
- sustain productive ecosystems
- retain future options
- retain economic opportunities.

These goals or desired outcomes are themselves interrelated and are consistent with the scientific basis for sustaining biological diversity – retention of the adaptive, generative capabilities of life (Bunnell 1998). The primary desired outcome is to reduce rates of species extinction. If that is achieved, the productivity and generative capabilities of ecosystems are likely to be sustained, and thus future options ("intergenerational equity" in international agreements) are likely to be retained. Economic opportunities are likewise related to the generative capacities of diverse populations and species (Bunnell 1997a, 1998). Broader international goals can be found in the convention, such as equitable sharing of resources derived from biodiversity, but most public concern is focused on the four goals noted; moreover, researchers have applicable knowledge relevant to them. Attaining these public goals should and can be the goal of management to sustain biological diversity. These goals are also consistent with the elements of ecosystem management reviewed by Grumbine (1994) and the broad vision of biodiversity offered by Rowe in Chapter 6.

(c) *Link monitoring to goals. Separate monitoring of goals from inventory of elements encompassed by biological diversity.* Measuring and cataloguing elements embraced by biodiversity is a different task from establishing goals and monitoring progress towards them. The scope of definitions of biological diversity has led managers and researchers to despair of knowing what the target should be (Bunnell 1994, 1998). Such despair can be justifiably reduced by focusing on the desired outcomes rather than on the myriad entities embraced by the term. As Pielou (1995) stated, "measuring is not a necessary preliminary to monitoring." For example, as Bunnell et al. (1991) and Namkoong (Chapter 3) note, detailed measurement of genetic diversity for the purposes of establishing a target is a misleading effort that will provide only a static "snapshot" resulting from processes known to be dynamic. Better to begin with publicly accepted goals such as those listed above.

Credible forest management cannot occur in the absence of quantifiable targets defined in terms of forest composition and spatial structure (question 15 as listed in Chapter 1 and discussed later in this chapter). These targets must reflect the dynamic nature of forest stands – that areas providing specific forest attributes will change in space with time (questions 8 to

10). Initially those targets will be informed guesses. Monitoring must therefore address both the desired attributes of forest structure (such as abundance of dead and dying trees) and the desired outcomes linked to those structures (such as productivity of cavity-users).

(d) *Focus research on the natural elements and processes that create or reduce diversity and on the assumptions that underlie management strategies.* Separate research from initial goal setting and focus it on these two broad topics. The scientific basis for maintaining biological diversity is not the wide range of elements noted within definitions, but variety or diversity itself (Wood 1997; Bunnell 1998). The definitions of McNeely et al. (1990) and the United Nations (1992) clarify this point. Neither definition claims that biodiversity consists of genes, species, ecosystems, physical structures, or ecological processes, but that it encompasses them and their variability. Biodiversity is most simply the differences among living entities, an attribute of life (Bunnell 1998). These differences within and among organisms permit continued adaptability, continued creation of biodiversity, sustained productivity in a changing environment, and thus sustained resources (Whittaker 1970; Frankel and Soulé 1981; Norton 1986; Naeem et al. 1994; Tilman and Downing 1994). This scientific basis for maintaining biological diversity, emphasizing sustained variation and adaptability rather than entities, connects directly with the public concerns noted for goal setting (Bunnell 1997a, 1998). Thus, research should address processes that create, sustain, or reduce diversity. For example, genetic research should focus on processes such as allele migration that sustain diversity, rather than on cataloguing genetic variation.

Research must also address assumptions that lead to choosing particular forest attributes as management targets. This second form of research addresses processes that create diversity but evaluates the linkage of those processes with specified forest features (for example, certain bird species are associated with hardwoods rather than conifers). Such research affirms the validity and effectiveness of strategic approaches and is thus an integral component of the process of monitoring the attainment of desired outcomes or goals.

(e) *Eliminate from goals components that are difficult to measure or define.* Acknowledge that it is impossible to measure all entities encompassed by the term *biodiversity* (e.g., Chapter 3 for genetic components; Chapter 4 for populations; Chapter 5 for communities and ecosystems). Accept surrogates to evaluate progress towards broad goals by using entities and states that can be monitored and that connect directly to societal concerns (see also "How Will We Know When We Get There?"). Direct research to the reliability of these surrogates and to the processes creating diversity or sustained productivity (described by Simberloff in Chapter 5).

(2) What Role Do Values and Perspectives Play in Defining the Goal?
Human values influence human undertakings. William Burch, longtime faculty member at the Yale University School of Forestry and Environmental Studies, summarized his experience by three central "laws" of resource management (as presented in Grumbine 1997):

- All resource allocation decisions are matters of political struggle (values) rather than technical facts.
- Resource management decisions are about use; therefore they are decisions about manipulating human behaviour rather than physical things.
- Resource managers, when confronted with social value decisions, will seek to convert them into technical decisions.

Given the pervasive role of values in our activities, it is helpful to group the effects of values at two broad scales (Chapter 1). The first is more general and strategic, frequently devolves to distinctions between intrinsic and instrumental values, and can influence broad choices in management actions (for example, preservation versus sustainable forestry) or the choices of questions pursued by individual scientists (for example, predictive models versus cataloguing detail). The second scale applies once a broad choice has been made and influences choices among tactics, such as which species or which area. Broader strategic values naturally influence values that govern tactical choices. No matter in which scale of values our decisions lie, when urgency is felt, values are expressed more fervently and may obscure information emerging from scientific processes. Scientific treatments of ecological or biological diversity also embody values (e.g., Franklin 1988; Pielou 1995; Chapter 5). The issue of values is large and challenging. Takacs (1996) describes the nature of the challenge somewhat glibly but accurately by noting that, at the pivotal conference on biological diversity in Washington, DC, in 1986 (Wilson and Peter 1988), "biodiversity" emerged as "biological diversity" minus the "logic." Despite its ubiquitous influence, I treat the issue of values briefly. Two steps are important: (a) accept that values play an important role, and (b) attempt to distinguish values from other information. These are discussed below.

(a) *Accept that values play an important role in specifying concerns embraced by biodiversity.* Values or perspectives play a role among the public, managers, researchers, and the larger public. There is, however, little coherence among these values. McPherson (1985, page 157) observed that "there is little agreement on how to value biological diversity, who should value it, and what dimensions of it should be valued." This condition derives from competing values, and as McPherson (1985) concludes, "a general approach to valuing biological diversity has eluded scholars and policy makers alike."

Nonetheless, numerous attempts have been made to define or list the values of biodiversity; Wood (1997) cites over 30 efforts and the problems associated with them.

As a guide to action, the assertion of intrinsic values in nature is problematic. Scherer (1990, page 4) expressed the problem this way: "Environmental ethicists have at most produced a theory of value. They have not produced a theory of action inferable from the former." Attempts to demonstrate ethical or aesthetic requirements for conserving biodiversity also founder (Chapter 5). From the pragmatic necessity of taking some action, we are driven to seek instrumental values, or reasons why we benefit from sustaining biological diversity (e.g., Bunnell 1997a, 1998; Wood 1997; Chapter 5). This outcome does not deny other values and is appropriate where actions are to be taken and resources used, as in managed forests. Four instrumental values strongly held by the public were noted under question 1 ("What is the goal?"). These often translate into concerns of managers or policy-makers through degree of emphasis. Foresters, for example, may choose to emphasize maintaining ecosystem productivity; policy-makers, future options; and wildlife managers or conservation biologists, rates of extinction.

Strategic perspectives also differ among researchers in two major ways. The first is a broad difference in world view. One group, which Sagoff (1993) characterizes as "demystifying the Great Chain of Being," searches for large, clockwork-like mechanisms through systems ecology. Their philosophical roots are in Plato's ideals; examples include MacArthur and Wilson's (1967) treatment of island biogeography, Levins's (1970) treatment of meta-populations, or writers on natural disturbance regimes and sustained yield theory (see also Botkin 1990). Another group, with roots in the approaches to cataloguing nature derived from Aristotle or John Ray, finds enough to admire and appreciate in nature's tiniest details, and includes individuals such as E.O. Wilson and S.J. Gould. Wilson, for example, is a self-confessed neophile: "I am a neophile, an inordinate lover of the new, of diversity for its own sake" (Wilson 1995, page 171).

Such broad differences may manifest themselves in a second related but distinctive way. Some individuals accept Ehrlich and Ehrlich's (1981) metaphor of "airplane rivets": the loss of any more species may represent the missing rivet that causes the plane to crash. Others adhere more to the "airplane passengers" metaphor derived from Walker's (1992) treatment of redundancy: many species present in a system are just going along for the ride and have nothing to do with flying the plane. Such differences in perspective influence most of what researchers do. In Chapter 5, Simberloff illustrates how alternative perspectives on the nature of communities or ecosystems (a "collection of species" or a "superorganism") guide considerations of monitoring. In Chapter 6, Rowe contrasts seeing a system from the inside and from the outside, and describes how biological and earth scientists arrive at different classification systems and different views of system organization. There is nothing wrong with different values or perspectives; indeed, they are a product of human nature. We benefit when we recognize that different management practices or research findings may have far deeper roots than the data on which they rely.

(b) *Attempt to distinguish values from other information.* A continuing challenge, particularly for policy-makers and policy approvers, is to distinguish human values from information derived by processes designed to be "truth preserving" (Popper 1959). By *information* I mean what people holding different values can agree on, because each would accept certain observations as evidence ("inter-subjectively testable" in Karl Popper's terminology [Popper 1959]). Science is a means, not an end; it is how we gain knowledge, not how we gain meaning. Thus, both science and policy benefit when the distinction between them is acknowledged (Bunnell and Dupuis 1995b). Science cannot contribute values and priorities to public decisions. Societal values will always be diverse and often change quickly. Attempts by scientists to be "truth preserving" in their methods yield "facts" that also change with time, but usually slowly. One result is that management practice and the education of practitioners to maintain societal values will always lag behind societal values (e.g., Binkley 1992; Namkoong 1993).

These two steps are important no matter the scale at which values enter, whether strategic decisions or tactical acts. I believe management strategies are more credible and effective when addressing instrumental values (see also Bunnell 1997a, 1998; Wood 1997). The dominant instrumental value, from which other values derive, is the generative capacity sustained by a diversity of living entities. Once a commitment has been made to sustaining biological diversity, values still influence tactical issues. Major tactical issues are captured in the four questions that follow.

Are All Features Equal?   As noted by Bunnell and Chan-McLeod in Chapter 1, the natural features included or emphasized in definitions of biological diversity are themselves wonderfully diverse, but not all features are equal. The fundamental basis for the values and concerns underlying biological diversity is genetic diversity. Other features of definitions of biological diversity (such as composition, structure, and function) are included primarily as attempts to capture and describe the range of genetic diversity that permits sustained generation of diversity, adaptability, and productivity. Despite this fundamental importance, genes have no capacity for independent function, and Namkoong (Chapter 3) and Merriam (Chapter 4) argue compellingly that species and their populations are the most practical units comprising biodiversity.

Higher in the scale of our discrimination, communities and ecosystems are inadequate features because they have ill-defined boundaries and are generally described as a collection of species, but the species themselves are constantly turning over (see Chapters 5 and 6). We define a community as a relatively stable collection of species (typically characterized by dominant species, omitting a long list of rare species in our discrimination). When species are lost from all communities, an entire package of genetic diversity is lost that cannot be reconstituted. Species thus become both a practical and reasonable surrogate for the fundamental basis of biological diversity.

This view is not universally held. Franklin (1993) argues for ecosystems (but see also challenges to Franklin's view by Tracy and Brussard [1994] and Wilcove [1994]). Rowe (Chapter 6) appears to promote landforms as the fundamental unit of biodiversity. Landforms have greatest utility at the planning stage, where, because of their physical nature, they can represent a truly nested hierarchy. Appearing to acknowledge this point, Rowe writes, "the land can be divided hierarchically into meaningful units from which organisms come and go" (Chapter 6). Organisms, however, are the essence of biological diversity and the justifiable focus of management attention. Although species are the most tractable unit, it is important to recognize that species are simply a more readily assessed surrogate for the more fundamental genetic diversity.

Other features or elements in definitions of biodiversity (such as populations, ecosystems, communities, and processes) represent groupings of individuals and species or consequences of their presence. See, for example, the definitions of the Forest Ecosystem Management Assessment Team (FEMAT 1993), Society of American Foresters (SAF 1991), or US Bureau of Land Management (Cooperrider 1991). Definitions created specifically to guide management commonly incorporate the concept of structures such as snags, downed wood, or seral stages (see Franklin 1988 and Franklin et al. 1989). This practice, however, results from efforts to create more readily measurable surrogates that reflect the presence of specific species or groups of species (Chapter 5). Delong (1996) and Bunnell (1997a) argue that operational goals should not be embedded in or confused with the definition of biological diversity. Whatever else comprises biodiversity, the central unit remains groups of individuals or subpopulations of species (Chapters 3 and 4).

Are All Species Equal?   All native species are equal in terms of defining goals but not in terms of emphasis in associated planning and monitoring. To be consistent with the societal goal of reducing rates of extinction, no species should be lost from a large area. Some species, however, are "weedy" or ubiquitous and require little management attention. It is for this reason that Pielou (1995) argues trenchantly against indices of ecological diversity, such as alpha diversity, as useful measures of biodiversity: alpha diversity may increase when a more specialized habitat is modified enough to accommodate a generalist species (see also Chapter 7). As human populations grow, a contentious undercurrent to this question will be more frequently confronted: are individual species analogous to the rivets holding an airplane together (Ehrlich and Ehrlich 1981) or to the unnecessary passengers in the plane (Walker 1992)?

Johnson et al. (1996) reviewed the relationships between species richness and the productivity and stability of ecosystems; currently, little unequivocal evidence supports either "rivets" or "passengers." Reflection on how we believe natural selection to work or on how communities are structured suggests that many species will be "passengers," at least locally, but we do not know which ones. Further, there is an unfortunate asymmetry in learning if we can reveal value only through loss. Given these observations, the current goal must include all native species. The allocation of limited resources to rescue current species, however, may benefit from some system of ranking (e.g., Yanchuk and Lester 1996). Simberloff (Chapter 5) notes how rankings can change with different values or scientific perspectives; Namkoong (Chapter 3) provides helpful suggestions for ranking that extend beyond genetic considerations.

Are All Places Equal? Not all places are equal, simply because there are better and poorer habitats for any given species (e.g., Chapter 5). Often, sites productive for the growth of trees are also richer in biological diversity and also more productive for other species (Bunnell and Dupuis 1995a). There are two implications: (1) sustaining biological diversity will be more difficult if efforts are restricted to less productive sites, and (2) the variety of life ensures that all sites contribute to biological diversity (the poorest site may contain unique species). Both issues should guide protected-area strategies. At a smaller scale (within an individual's home range), it is important to recognize that individual species obtain different life requisites in habitats that differ in any single measure of "quality." This phenomenon, coupled with the incessant dynamics of a forest and the variety of life, reveals that movement of individuals or disseminules across a forested landscape is critical to ensure that biological diversity is maintained.

A related issue is that of maintaining a species across its current range. There are two aspects. The first relates to the local disappearance of individuals within a large area, which may disconnect populations and expose them to greater likelihood of extinction (see Chapter 4 and references). This aspect influences monitoring and suggests the usefulness of presence/absence checklists across large planning units specifically designed to detect growing holes in the range or shrinkage from the periphery (e.g., Scientific Panel[1] 1995).

1 Scientific Panel for Sustainable Forest Practices in Clayoquot Sound.

The second aspect concerns the amount of effort that should be expended at the periphery of a species' range. The answer to policy-makers is the same for either alleles or species: effort should be expended. As Namkoong notes in Chapter 3, "such simple techniques as targeting samples ... [or retaining] populational outliers or ... distributional extrema will increase [genetic] diversity with or without divergent selection." Scudder (1989) and Lesica and Allendorf (1995) provide similar arguments for maintaining peripheral populations. Lesica and Allendorf attempt to determine under what circumstances peripheral populations are beneficial to the evolutionary process. Peripheral populations are important because some species have declined dramatically by a gradual whittling away at the periphery of their range. This process is especially evident among large carnivores such as cougars (Chapter 7) or grizzly bears (Wielgus and Bunnell 1994).

Combined, these observations have implications for both managers and policy-makers and emphasize the need to maintain movement of individuals and thus genes. The manager must plan activities within large planning units in such a way that they do not greatly restrict movement. Policy-makers can assign the maintenance of particular species to specific areas or political jurisdictions only for the most localized species. Localized preservation may serve to maintain species, but it will not maintain inherent diversity. Hidden in this observation is a fundamental paradox related to scales. Isolation through restriction of movement leads to localized extinction in the short term (Newmark 1995; Chapter 4 and references). Over the much longer term, isolation appears to be the major mechanism of speciation and generation of diversity (Simpson 1944; Mayr 1963). This paradox emphasizes the need to separate the scientific and operational definitions of biodiversity. The long-term role of isolation and subsequent speciation is ill defined; the short-term effects of isolation can reduce diversity. Current management must focus on maintaining connectivity among populations while research grapples with the paradox.

What Risks Should We Take?   As few as informed management permits. Agreements emerging from the United Nations Convention on Biological Diversity of 1992 explicitly address risk by arguing for retaining future options or intergenerational equity (Bunnell 1997a). Intergenerational equity charges us with being good ancestors and not taking undue risks with current resources. Risks are issues of policy because, with the array of values that society asks of forests, maintaining some values is in direct conflict with maintaining others (e.g., Chapter 1). Noss (Chapter 7) argues for a cautious approach. Maini (Chapter 2) and Merriam (Chapter 4) explicitly invoke the "precautionary principle." This principle can be summarized simply: act cautiously and make subsequent adjustments based on the application of methods tested and found successful in similar environments; modify actions appropriately by diligent monitoring of responses.

Research is affected by the precautionary principle in three ways. First is the need for researchers to be good ancestors by focusing their creativity on questions that will most effectively inform management. Bunnell and Dupuis (1994, 1995b) argue that trends in the published literature suggest that researchers are not fully exploiting available opportunities. Bunnell et al. (1997) and Bunnell and Huggard (1998) note that the "cerebral anarchy" of many researchers impedes the provision of information that is useful to practitioners. Second is the compelling need for rigorously applied adaptive management to help inform managers and policy-makers in an ongoing way (see question 17). Third is the potential for research to derive minimum population sizes that provide low probabilities of extinction (Soulé 1987).

Noss (Chapter 7) provides examples of the approach. The approach, however, is useful only for a small portion of the entities that make up biological diversity, and is better directed towards assessing tactics for species known to be endangered (e.g., Chapter 3). Moreover, most population viability models (reviewed in Burgman et al. 1993) have considered the special case in which ecological factors have no impact on vital demographic rates. The literature on this topic is currently very active. Relevant relationships (e.g., Avise 1995) and generalities (e.g., Frankham 1995) are being derived, and the concepts may soon provide more general tools.

(3) Over What Area and Time Period Should the Goal Be Defined?
Previous chapters consistently emphasize that processes engendering diversity extend across spatial and temporal scales. Namkoong (Chapter 3) notes that genetic diversity is a necessary but insufficient condition for species survival. Merriam (Chapter 4) observes that even though species are important, the gains and losses of diversity are enacted through population processes. A fundamental process sustaining both genetic and species diversity is movement of individuals among habitats and ecosystems. Rowe (Chapter 6) notes that "the bio-ecologist's ecosystem tends to be a conceptual device relatively elastic in its space/time dimensions," and argues for big areas and long time periods, as does Noss in Chapter 7. Given the inevitable lack of discrete boundaries in systems composed of organisms (Chapters 5 and 7), four steps are important (see also question 10): (a) accept a broadly defined target, (b) use units based on landforms, (c) use areas and periods that make biological sense, and (d) invoke hierarchical planning. These are discussed below.

(a) *Accept a broadly defined target* (see also question 9). The goal must be defined within an envelope of variability and not as a static entity. The envelope of natural variability, including disturbances such as natural mass wasting, floods, and forest fires, can serve as an interim goal. Society almost certainly will seek to make that envelope more restrictive than nature allows (see question 12). No matter how broad the target, it cannot be static; there is nothing static about nature. With no human intervention, species distributions within an area will change in response to natural disturbance regimes and natural succession, even over the short term (MacArthur and Wilson 1967; Bunnell 1990, 1992, 1995; Chapters 5 and 7). Over the longer term, changes occur through physical processes that modify landscapes and through biological processes that constitute natural selection (Chapter 6). Indeed, Noss (Chapter 7) argues that "Nature is too variable and unpredictable for us to specify a future condition" (or specific target), and instead suggests that goal setting identify only the desired direction of change.

There is an important caveat: if people accept that nature is too variable for us to specify any desired future condition, there may be "paralysis by complexity" (e.g., Bunnell 1994, 1998) and little effective management. By "broadly defined target" I mean that the forest cannot be the same from one year to the next. Specific targets must recognize and incorporate the naturally dynamic processes of forests. One approach is to specify the minimum desired levels or thresholds of particular attributes such as amounts of various seral stages, older forests, snags, or downed wood. A current problem is that such targets represent a strategic view of desired forest structure, whereas most public concern is focused on tactics such as the method of harvest. Policy-makers, managers, researchers, and the public must focus first on the desired outcomes, then on strategy, and lastly on the appropriate tactics.

(b) *Use units based on landforms.* Rowe (Chapter 6) lists five compelling reasons why landforms are useful, even fundamental, units. Other reasons are presented throughout his text. The most important of these may be that landform features "anchor" and "control" the ecological relationships of terrestrial ecosystems, and that landforms are accessible to mapping and inventory in ways that entities distinguished by their moving parts (such as communities) are not.

(c) *Use areas and periods that make biological sense (for example, large enough for a sustained forest management plan, at least one rotation in length).* Any achievable goal must treat areas that are large enough so that known entities embraced by biological diversity are not absent long enough to alter the sustained generative capacity of the area (entities such as species are always being lost and replaced within small areas). Sustained generative capacity is recognized by societal goals of retaining future options and maintaining productive ecosystems. Truly long-term processes that fundamentally change biodiversity (for example, natural selection) elude planning processes and cannot be incorporated into a goal – at least not directly. Appropriate management consists of short-term steering towards some long-term goal. Natural selection will help generate future options in the long term, provided variety is retained in the short term.

For most forest types, both the ideal upper size and the ideal time horizon are impractical (see Chapter 7). Goals and plans should be defined over smaller areas and shorter time periods in a manner that we believe will sustain important events occurring at larger and longer scales. This objective can be accommodated by hierarchical planning (e.g., Scientific Panel 1995; Chapters 6 and 7). Specific sizes or rotation lengths appropriate for sustainable harvest will vary with forest type but are likely to be on the order of 200,000 to 400,000 ha and 70 to 120 years (see also question 9).

(d) *Invoke hierarchical planning.* Hierarchical planning explicitly recognizes that elements embraced by the term *biodiversity* cross a range of spatial and temporal scales. As Bunnell and Chan-McLeod (Chapter 1), Namkoong (Chapter 3), Merriam (Chapter 4), Rowe (Chapter 6), and Noss (Chapter 7) all note, no single scale is appropriate. A hierarchy permits the linking of entities and processes with appropriate scales. The approach is explicitly "top down" and, for biological diversity, recognizes that a major goal is not to lose species from large areas (about 200,000 ha or more). The significant point is not the actual scales in the hierarchy but the hierarchy itself. Rowe (Chapter 6) discusses the advantages of hierarchical classification for inventory and provides examples for Canada and British Columbia. Noss (Chapter 7) provides an example focused on one species (coastal sage shrub) and its associates; the Scientific Panel for Sustainable Forest Practices in Clayoquot Sound (Clayoquot Scientific Panel) provides a more generic approach embracing all forest values (Scientific Panel 1995).

It is important to note that although a hierarchical approach is useful in planning, it frequently fails as a research strategy. The usefulness of a hierarchical approach to planning is manifested in physical systems where some smaller systems are indeed functionally nested within larger systems (for example, smaller watersheds within larger watersheds). It is from the basis of physical landforms that Rowe (Chapter 6) presents his arguments. The structure and relationships of the biota are rarely functionally nested, so fine-scale research rarely scales back up to the broad scales for which managers seek information (Bunnell and Huggard 1998). One consequence is that much fine-scale research overlooks the broad-scale effects of physical setting on research findings, such as the influence of underlying geology or stream gradient on the response of tailed frogs to forest practices (Levin 1992; Bunnell et al. 1997). Rowe (Chapter 6) alludes to this difficulty: "the land can be divided hierarchically into meaningful units from which organisms come and go." Organisms, however, are the essence of biodiversity. An important point is that while a hierarchical approach to planning is helpful, especially in the sense of short-term steering, it can hinder the understanding of processes sustaining biodiversity (many are not functionally nested). Without understanding, management actions will likely have unanticipated consequences.

(4) What Is Ecosystem Integrity?

Ecosystem integrity is a bridging concept that merges the observations of science with the values and judgments that make science a human endeavour. Its usefulness to goal setting is that it links the values we seek to actions that sustain biodiversity, such as maintaining healthy or productive ecosystems (Bunnell 1997a).

Either implicitly or explicitly, writers on biodiversity equate maintaining biological diversity with maintaining ecosystem integrity or health (see, for example, Chapter 7). The inclusion of "processes," "structures," and "functions" in some definitions of biological diversity (e.g., Franklin 1988; SAF 1991; FEMAT 1993) reflects efforts to define an integral whole. Generally the term *ecosystem integrity* is meant to signify functioning, self-sustaining systems undergoing no systematic changes as a result of human-induced manipulations (Scientific Panel 1995). Like *biological diversity*, it is not a scientific term but rather one of what Ehrenfeld (1992) called "bridging concepts" – concepts that connect a scientific concept about the state or properties of a system with a social value about the normative or desired state. For example, when referring to our bodies, "health" incorporates human values that are not amenable to strictly scientific measurement. "Health" can, however, be a useful reference concept for identifying the *stresses* (another bridging concept) to which bodies, watersheds, or ecosystems are subjected (Scientific Panel 1995).

In terms of goal setting, identifying symptoms of ecosystem stress and response to stress might lead to a set of diagnostic principles for assessing ecosystem state (Schaeffer et al. 1988). It is difficult, however, to set a target state for ecosystems subject to natural disturbances (Bunnell et al. 1993; Cumming et al. 1996); such systems span a wide range of states, all equally healthy (Bunnell et al. 1997). Scientific methods can describe changes to a system in response to a disturbance and can determine causal mechanisms for most major disturbances (including human interventions). The question of whether or not a system is "healthy" or has integrity is a question of value and interpretation. This situation is appropriate. The term *biodiversity* itself reflects values (questions 1 and 2). Moreover, managing forests or any feature of the natural environment entails recognizing and incorporating human objectives for the system, even when a deliberate attempt is made to ground management firmly in scientific principles. It is unlikely that the term *ecosystem integrity* includes significantly more than is embraced by the four societal concerns noted under question 1.

(5) Can Indices of Ecological Diversity Help Define Goals?
Biological diversity is not ecological diversity. Although widely used, indices of ecological diversity contribute little to goal setting.

Several definitions of biodiversity (OTA 1987; NSF 1989; Reid and Miller 1989; Erwin 1991; McAllister 1991; Spellerberg 1992; Art 1993) explicitly incorporate components of ecological diversity (species and their relative frequency). The reason seems to be that many definitions have simply rephrased that of the US Office of Technology Assessment (OTA) (see review of Delong [1996]) or are a search for some measurable attribute. There are compelling reasons to avoid using diversity indices in specifying management goals. Several authors have emphasized this point (Pielou 1995; Bunnell 1998; Simberloff, Chapter 5). Bunnell (1998) summarized the arguments briefly:

- Diversity indices treat each species as equal. Except for turnover, measurements of ecological diversity would assess the replacement of a specialized species by a generalist through habitat modification as an unchanged state (alpha diversity would remain identical). This is contrary to societal concern for keeping all species. Although all native species are equal in terms of initial goal setting (question 2), this does not mean that species richness or numbers are the target if specialists are replaced by more ubiquitous generalists.[2]
- Diversity indices are commonly applied to narrowly defined groups of organisms, whereas biodiversity is intended to be a property of the biosphere or an attribute of life.
- Ecological diversity is a theoretical subject, not an operational goal (see Chapter 5). For this reason the indices have usually been applied to narrowly defined groups of organisms and have been subjected to decades of debate about their interpretation (see reviews in Eberhardt 1969; Hurlbert 1971; Peet 1974; Pielou 1975).
- Diversity indices commonly emphasize species evenness or equitability, which is an unnatural condition. High equitability in species abundance is theoretically interesting and intuitively more diverse (see Chapter 5), so indices of ecological diversity give greater weight to the more evenly distributed groups of species (Bunnell 1998). Equitable distributions of species, however, are extremely rare (see review in Magurran 1988).

Indices of ecological diversity are not valueless in setting goals for biological diversity. Despite 55 years of argument over their theoretical meaning, they may help us think about processes underlying attainable goals.

---

2 Noss (Chapter 7) appears to argue for diversity indices, but the example he provides offers little useful information without a knowledge of individual species.

For example, it is clear from examining indices that in any given community most species are relatively rare (Magurran 1988), which suggests that many are "passengers" rather than "rivets" or "pilots." Similarly, the concepts of alpha, beta, and gamma diversity appear to illustrate potential management tactics by revealing natural rates of species turnover among communities (but see question 14).

(6) How Different Does an Entity Have to Be Before It Contributes to Biodiversity?
Species and easily recognizable subspecies are useful divisions of variation. Subpopulations of these are important in monitoring.

The force and content of the term *biodiversity* derives from the fact that it reflects differences within a collection of living things. These differences yield the generative, adaptive, and productive capacity of biodiversity. The Convention on Biological Diversity explicitly distinguishes between the entities embraced by biological diversity ("biological resources") and biological diversity itself. The convention defines biological diversity as "the variability among living organisms from all sources" (an attribute of life), thus different from biological resources, or living entities, which include "genetic resources, organisms or parts thereof, populations, or any other biotic component of ecosystems" (United Nations 1992).

The question that arises is, how different does something have to be before it contributes to diversity? In terms of goal setting, what range of differences should the goal recognize? At this fundamental level, the scientific content of the term *biological diversity* is unclear. The natural world creates both the problem and the solution.

The problem occurs among vertebrates because most species reproduce sexually, thus each individual has a unique mix of genetic material (Selander 1976) and contributes diversity. Elimination of any individual reduces genetic diversity. This problem is seemingly worsened because, even if some alleles were more valuable than others in generating subsequent diversity, combinations of alleles are usually invisible from the outside.

Nature also suggests the solution. Like community composition, genetic diversity is an ever-changing, ensemble property (Chapter 3). The generative contributions of genes, however, occur through population processes within formed or evolving species (Chapter 4). Within large groups – such as vertebrates, insects, or fungi – boundaries between species are variably discrete, but each species generally represents a complete, self-generating, and variable ensemble. Species are thus the largest tractable units of variety – communities and ecosystems change more rapidly through species turnover. Phenologically distinct subspecies represent visibly different ensembles (Bunnell and Williams 1980). The maintenance of generative capacity within species or subspecies, however, depends on the movement of individuals. If movement between infrequently connected subpopulations becomes restricted, isolated subpopulations will either disappear or diverge from the other subpopulations (depending on their size). More or less separable subpopulations (see Chapter 4) thus become a feature of monitoring. We arrive at the interim construct that species or visible subspecies represent appropriate differences, but that monitoring must consider the subpopulations that make up a species and its range.

(7) Is Viability (Large Tracts) or Representation (Several Smaller Tracts) the Dominant Goal of Protected Areas?
The range of life forms encompassed by biodiversity prevents a simple choice and emphasizes the importance of maintaining conditions favourable to movement or dispersal within managed forests.

Protected areas pertain to goal setting because they are assumed to be fixed, reliable vessels for sustaining biodiversity. Human activities then alter the managed forest surrounding them. Nature, however, alters the protected areas even when human presence is negligible. Rowe (Chapter 6) argues for representation but does not engage the SLOSS argument ("single large or several small" reserves); neither does Namkoong's (Chapter 3) argument for the utility of small populations ("extrema," as at range peripheries). The argument will not go away, however, because the answer inevitably depends on chance outcomes for which probabilities are poorly known and possibly changing, and on outcomes that are very different across life forms.

Current research has documented little more than the concept that larger areas are less likely to show loss of species than smaller areas (Newmark 1995 and references therein), and that smaller, representative areas appear useful for specific plants, arthropods, and other less mobile species (e.g., Ryti 1992; Prendergast et al. 1993). Concentrating preservation efforts in a single large area makes that single area vulnerable to unpredictable events (Chapter 7). Moreover, areas of concentrated richness for some taxa are poorly representative for others (Prendergast et al. 1993). An equally contentious issue in goal setting is "how much is enough?" Rowe (Chapter 6) reviews the work of researchers, including Noss, who have argued that 33-50% of specific regions should be committed to protected areas. In each case these researchers advocated a form of zoning that would fully commit some lands to resource production for human use.

Policy-makers should recognize the following:

- Protected areas are passive, optimistic (unless very large) ways of maintaining biodiversity (Chapter 3).
- Any percentage target is necessarily arbitrary, although more is better.
- A universally fixed percentage (for example, 12% protected) is even more arbitrary and does not reflect the heterogeneous distribution of diversity.

Parks experiencing significant human use may not be helpful in sustaining biological diversity (Gibeau 1995; Knight and Gutzwiller 1995).

• Zoning the intensity of resource extraction likely provides greater flexibility than attempting to meet all values everywhere (Bunnell 1996; Binkley 1997).

• Any target for protected areas is a useful but unreliable part of goal setting (chance events will modify usefulness).

The last point has implications for management. Protected areas are useful because they designate areas of the landscape where changes other than park development will not occur through direct human intervention (indirect effects, such as climate change, are pervasive [Hebda 1997]). Although a form of passive management, their establishment reduces uncertainty in forest planning and allows practices to focus on a managed forest matrix in which conditions favourable to movement can be maintained for many species (Bunnell et al. 1997). The chance events themselves are functions of the natural disturbance regime and can guide practices. For example, deliberate management actions to maintain connectivity or movement among subpopulations within the managed matrix are more important where natural disturbances are small and infrequent (Bunnell 1995). In short, there is no tidy answer to this question, nor should we expect one, given the diversity of nature. It would be helpful, however, if some sizable protected area system were created and surrounding management were reconciled to and coordinated with its existence (e.g., Ohmann et al. 1994; Bunnell et al. 1997).

How Do We Get There?
The second broad question is tactical and defines the kinds of activities that will help us attain the specified goal.

(8) How Do We Develop Strategy or Tactics When the Target Is Poorly Defined and the Scientific Information Is Largely Incomplete?
There are five major steps. The most important is not to confuse goals with details.

• *Focus on goals, not details.* Broad goals such as those described for question 1 are sufficient to plan actions. Focusing on the scope of life or detail embraced by biological diversity leads to paralysis by complexity (Bunnell 1994, 1998).

• *Employ large enough planning units and long enough planning horizons that the larger-scale processes sustaining diversity are accommodated.* These are at least 200,000 ha and one rotation length (see question 10). Do not search for a single relevant scale; there is none (Levin 1992; Bunnell and Huggard

1998). Hierarchical planning assists in linking scales at which processes occur to the appropriate planning scale (e.g., Scientific Panel 1995).

• *Fit the tactics to the goals.* Species are the most tractable unit of emphasis (questions 2 and 6). Namkoong (Chapter 3) provides a summary of appropriate tactics based on current knowledge and perceived function or use of the species. A continuing problem is public concern over tactics (such as clearcutting) while the desired outcomes remain ill defined. The focus must be first on the desired outcomes, then on strategy, and lastly on the appropriate tactics (see question 3).

• *Do not confuse inventory with monitoring.* Monitoring must fit the goals (Chapter 5). It is impossible to monitor all entities embraced by biodiversity, nor is it necessary (Pielou 1995; see also question 1). Bunnell (1997a) offers suggestions on how monitoring can be focused on the major outcomes sought; the Scientific Panel (1995) provides a more complete approach to monitoring.

• *Practice adaptive management.* Most simply, adaptive management consists of changing management when monitoring reveals undesirable outcomes. The utility of adaptive management is treated under question 17. Here it is important to note that adaptive management is similar to statistical forecasting in that it need not rely on specific, mechanistic understanding (Chapter 5). Bunnell (1989) described a dichotomy in the purposes of models – prediction versus understanding – and noted that these purposes or goals can proceed on separate tracks. He argued that predicting management outcomes need not wait for scientific understanding, but can still exploit adaptive management. Although challenged by some modellers (e.g., Hengeveld 1992), this view is supported by others (e.g., Peters 1991; Shrader-Frechette and McCoy 1993; Simberloff, Chapter 5).

(9) Given That the Target Moves, How Can We Develop Appropriate Tactics?
Tactics should be designed to maintain variability within broadly defined limits.

Everything about nature is in flux. Genetic diversity shifts continuously through reproduction and mortality. Species compete with each other and move into and out of habitats created by natural disturbance regimes. Communities gain and lose species, and even landforms change their shape, albeit slowly. The greatest value of biological diversity, or variability, is that this diversity itself generates productivity and further diversity. Public and scientific concerns are primarily about loss: loss of species, loss of productivity, loss of future options, loss of economic opportunities. Tactics should not simply accept a moving target but be deliberately designed to create an ever-changing landscape (as long as all necessary structures such as snags and downed wood are present). *The most destructive option would be to do everything the same way everywhere.*

There are three large implications of the tactics proposed. First, the entire forest area cannot be committed to intensive fibre production. This would not only make the reliant social infrastructure vulnerable to unexpected variability but would limit options to create the variability upon which biological diversity depends. Second, because the limits where losses occur are ill defined, caution and the precautionary principle become important guides to planning (see question 2). For the same reason, adaptive approaches to management are important. Third, because losses (for example, of species from a local area) occur naturally, it is less important for research to attempt to define a natural disturbance regime than it is to assess thresholds that increase loss rates (Bunnell 1997b; also questions 4 and 12).

(10) What Are Appropriate Planning Units and Time Horizons?
The basic unit should be defined by landform and should be at least 200,000 ha in area, planned over one rotation length.

This recommendation assumes that protected areas are planned over much larger areas. Most entities mentioned in definitions of biological diversity (populations, communities, ecosystems) naturally have constantly shifting boundaries and interchangeable parts – genes, subpopulations, species (Chapters 3, 4, 5). Landforms such as watersheds have much more permanent, recognizable boundaries that delineate actual space or location (Chapter 6). Actual location matters not just for planning but because forestry practices modify features at real places, and real space is important in processes, such as fragmentation or isolation, potentially influencing the flora and fauna (Bunnell and Kremsater 1994; Chapters 4 and 6). Because they constrain movements of organisms, air, and water (Chapter 6), landforms serve as natural boundaries regardless of administrative boundaries. Obvious natural units are individual watersheds, or aggregations of smaller watersheds, large enough to accommodate both natural variation (question 9) and planning horizons matching forest practices.

Rowe (Chapter 6) argues for land-use planning on a grand scale and with a long time horizon; he is supported by Noss (Chapter 7). Such a visionary approach would greatly aid the maintenance of biodiversity, but it is equally important to address immediate biological and operational constraints. Forest-dwelling species inhabit and are sometimes largely restricted to specific structural or seral stages, from recent disturbance through old growth (Hagar et al. 1995; Bunnell et al. 1997). The planning unit must be large enough to include all seral stages as they are generated through one complete rotation, so that amounts, distribution, and location of seral stages can be evaluated. Although the size of appropriate planning units will vary with forest type, they are unlikely to be much less than 200,000 ha

(Chapter 7). Smaller areas can be planned, but they are unlikely to represent a *sustainable* forest management regime because they are economically dependent on a larger area and vulnerable to changes there.

The minimum time horizon for planning is a rotation, which is also variable among forest types. This minimum is set both by the requirement that the full range of seral stages be evaluated and by the need to assess the economic viability of the approach. Currently, planning over such large areas and long time periods is possible only through computer-based spatially explicit decision support tools (e.g., Daust and Bunnell 1994). The achievable upper size is currently set by the efficacy of these tools. Given the number of computations necessary to assess the responses of biological diversity to alternative forest management plans, it is difficult to adequately assess areas larger than about 200,000 ha.

A further limitation determines the appropriate length of the planning period. Existing planning tools that attempt to incorporate biological diversity in forested systems (e.g., Hansen et al. 1993; Daust et al. 1994; Daust and Bunnell 1994) necessarily incorporate many poorly understood relationships. Such tools are precise, but are usually precisely wrong. The longer the planning horizon, the greater the cumulative inaccuracy of small errors (Bunnell 1989). Given this observation, it is unwise to attempt detailed planning much beyond 20 years, and most projections through an entire rotation will be unreliable. Our current inability to project biological relationships accurately over an entire rotation is not a dangerous liability provided we incorporate some monitoring and adaptive management. It is important that we acknowledge current management as short-term steering towards long-term goals, and change practices quickly when they are found inappropriate. These comments apply to the basic planning unit over which sustainable forest management can be considered. Management efforts focused on specific elements of biological diversity naturally exploit other spatial scales (see Chapter 3).

Policy needs to include the direction of efforts to reconcile aggregations of watersheds with arbitrary administrative boundaries and the facilitation of interagency cooperation. The importance of the latter in attaining conservation goals cannot be overemphasized (Slocombe 1993; Grumbine 1994; Ohmann et al. 1994; Galindo-Leal and Bunnell 1995).

(11) What Are the Relative Roles of Protected Areas and Managed Forests in Biological Conservation?
This is a policy question with no single correct answer. The first step is to clarify the goals of the protected-area strategy.

One unequivocal role of protected areas is to provide baseline data to aid in monitoring (question 16). It is less clear how protected areas should contribute to maintaining biological diversity. For example, a major goal of a protected-area strategy for forested areas could be to maintain late succes-sional species. This appears to have occurred in the American Pacific North-west. There, very little timber harvest is occurring on forested federal lands, which are de facto protected areas. Relieved of the challenge to maintain late successional species, state and private lands are managed more inten-sively and contribute more dollars per hectare to social infrastructure (Bunnell 1996). The approach results in an effective but implicit zoning of intensity of forest management and, were it explicit, would facilitate planning, re-search, and management tactics (Bunnell et al. 1997).

Planning would be facilitated by clearly specifying the desired products from each kind of area. Research could then focus on (1) the capability of protected areas of different size and distribution to maintain late succes-sional species, and (2) specifying elements of biodiversity that must be pro-tected from human intrusion. Management would be facilitated both by having clear objectives and by reduced dissonance among competing val-ues. If, for example, sufficient protected areas were designated, most man-aged forests would not require long rotations incompatible with economic contributions to social infrastructure. Moreover, appropriate management tactics could be selected for each area (see Figure 3.2).

The goals of protected-area strategies are usually broader than in the above example. They include the goal of representation (question 7), and often fail to acknowledge that some species can be maintained within managed forests. When the goal is diffuse, associated research is also diffuse. Because all species are currently encompassed by mandates of most protected-area strategies, research findings contribute little beyond the points of question 7 (Bunnell 1996). When protected areas must "preserve" all of biological diversity and when managed forests "sustain" all of biological diversity, the competition between the two for physical space is intense (Chapter 1). It is probable that where considerable old growth remains, assigning responsi-bility for late successional species to protected areas would increase produc-tivity on managed forests so that more area could be protected without reducing overall economic contributions to social infrastructure.

(12) Should Managed-Forest Mosaics Mimic Those Generated by Natural Disturbance Regimes?
No. It is an impossible task and socially unacceptable. The task is impossible for three reasons:

- Natural mosaics change unpredictably and are a moving target (Bunnell 1992; Swanson et al. 1993; Bunnell and Kremsater 1994).
- We have no compelling argument for any particular period over which to define the target, as natural disturbances such as fire, insects, or ice ages

show different periodicities at different times (Clark 1990; Frehlich and Reich 1995; Cumming et al. 1996).

• Such a target invokes an untenable philosophical argument that humans are unnatural (Haila and Levins 1995; Shrader-Frechette and McCoy 1995).

Mimicking natural disturbance regimes is socially unacceptable because some completely natural disturbances are simply too large, rapid, and disruptive to incorporate into any planning process. Society will not accept a 20,000 ha clearcut implemented over a few weeks.

Nevertheless, the natural disturbance regime as a model for forest management has been propelled by genuine scientific and public concerns. The issues are similar to those surrounding biological diversity in that there is no clearly defined target and the underlying concerns are not explicitly addressed by the term that encompasses them. Scientists are concerned for two broad reasons. First, recent forest practices are reducing amounts of specific forest attributes (for example, snags and downed wood) upon which elements of biological diversity depend (e.g., Cline et al. 1980; Bunnell and Allaye-Chan 1984; Spies et al. 1988). Second, forest management has distributed these practices through space and time in a manner that threatens some species (e.g., Angelstam and Mikusinski 1994 and references therein). Most public concern derives from three sources: (1) scientific concerns legitimize public fears, (2) "natural" seems somehow better than "unnatural," and (3) the public does not like the way some forest practices look.

Natural disturbance regimes are championed primarily because of fears of losing species or productivity (see question 1). Given our inadequate understanding of most species, the best way of conserving them all is to ensure that systems retain their overall structure and function (Walker 1995). That structure and function is a product of the natural disturbance regime. Natural disturbance regimes thus became a goal that seemed simpler to achieve than sustaining biological diversity. For boreal and other forests naturally subject to extensive disturbance, however, it is most unlikely that the public will accept a truly natural disturbance pattern. The publicly borne costs of fire suppression alone confirm this. One corollary is that truly natural appearances are also socially unacceptable. The effective compromise is to embrace the concerns underlying a commitment to sustain biological diversity (question 1) and ensure that structures generated by natural disturbances that ameliorate such concerns are maintained in managed forests. The Biodiversity Guidebook of the Forest Practices Code of British Columbia (BC Forest Service 1995) appears to have sought this compromise. It has been less successful in reaching the compromise for two reasons: (1) the number of guidelines is large and, when combined, produce unintended consequences; and (2) efforts have become focused on defining a natural disturbance regime instead of determining what elements of it are valued for specific reasons (Bunnell 1997b).

(13) Can We Assess the Relative Impacts of Efforts Devoted to Stand-Level and Forest-Level Activities?
No. The degree to which forest-level effects are the "error term" or cumulative result of stand-level effects is unclear.

Tactically, the question is important because the answer determines the amount of effort and funds that should be committed to stand- or treatment-level practices (such as feathering clearcut edges) as opposed to forest-level effects (such as movement corridors). Moreover, knowing the scale at which practices most affect biological diversity would guide monitoring approaches.

There are implications for planning and research as well. If forest-level effects are simply the sum of stand-level effects, then planning tools need only sum the effects of edges or seral patch size, for example, to project the consequences of forest management. Research could be restricted to such effects and ignore potential emergent phenomena extending over larger areas. Actually, most research has done that simply because larger-scale studies are logistically difficult. Many studies that seemingly address larger scales, such as corridors for movement, do not do so (see critiques in Hobbs 1992 and Simberloff et al. 1992). As a result, the implicit assumption that stand-level practices and studies are sufficient to guide forest-level outcomes remains untested.

Four features are apparent in the current literature:

- For montane forests of the Pacific Northwest there is no evidence of forest-level effects such as fragmentation on birds or mammals (Bunnell et al. 1997). This observation suggests that focusing on treatment-level or stand-level effects in the short term may not be detrimental.
- Researchers have contributed relatively little to the resolution of cross-scale phenomena because they often emphasize precision and looking inward instead of generality and looking outward (Bunnell and Huggard 1998). These authors also report evidence that both fine-scale patterning and broad-scale patterning are important even though they do not fit tidily under current concerns of edge effects and fragmentation. Their analyses confirm those of Bunnell et al. (1997) that the most important action to avoid is doing the same thing everywhere.
- Large areas and long time periods must enter planning because important processes (such as widely ranging or migratory species, or hydrological regimes) extend over large areas, and cumulative effects (such as cumulative rates of cut, future snags, and downed wood) are important.

• We lack conceptual models that integrate genetic, population, community, and landscape scales of management with compositional, structural, and functional aspects of biological diversity. Such models represent a clear research need, but their development is hindered by a commitment to precision instead of cross-scale patterning (Levin 1992; Bunnell and Huggard 1998).

(14) Can Diversity Indices Help Define Tactics?
Very little, and not as often perceived.

The potentially misleading nature of diversity indices that quantify the average rarity of species in a community and their lack of usefulness in goal setting have been noted (question 5; Chapter 5). Indices such as alpha, beta, and gamma diversity are more often related to management tactics. Most simply, alpha diversity is the number of species in a small, relatively homogeneous area (such as a stand), and gamma diversity is the number of species in a much larger, less homogeneous area (such as a biogeoclimatic zone or biome). There are at least six different ways of measuring beta diversity (Magurran 1988), but the major distinction is that beta diversity is not a number but a ratio quantifying the turnover of species between one community or stand and another.

Within a large, contiguous old-growth forest, samples of various parts would likely reveal a low rate of difference among species lists (little beta diversity). Initial forest harvest would likely permit new species to enter the recently disturbed area, and beta diversity would increase among portions of the forest, as would gamma diversity for the entire tract. Some workers (e.g., Namkoong 1991; Samson and Knopf 1993) have noted that this increase in beta diversity argues for creating a mix of age classes or otherwise "artificially" increasing diversity in patches of forest. For other workers, such notions are contrary to the content or meaning of the term "biological diversity" because they would treat "weedy" as well as "sensitive" species equally; Pielou (1995), for example, considers Namkoong's arguments "an aberration."

The problem is that each author is, to some degree, correct. Natural disturbances continuously alter forest structure. Deliberately altering a forest's structure across a landscape will permit more diversity *provided* enough of all structures are present and appropriately distributed across the landscape. The insight these indices offer for management tactics, however, is not in revealing how forest harvest can increase species richness. The indices help by revealing just how large an area is required to set an operational goal and to provide a useful planning unit (see question 3), how heterogeneous the communities of seemingly uniform landscapes really are (e.g., Lapin and Barnes 1995), and how important it is to maintain both broad-scale and fine-scale heterogeneity if all species are to be accommodated (Bunnell et al. 1997; Bunnell and Huggard 1998).

How Will We Know When We Get There?
This third broad question is evaluative and acknowledges that some measure of success is necessary. The major issues or difficulties arise largely from those noted with goal setting and appropriate tactics. They can be summarized briefly.

(15) What Do We Monitor?
Monitor elements of nature that link directly to public goals encompassed by the term *biodiversity*.

Initially we must monitor progress towards the desired outcomes that led to a commitment to maintaining biological diversity (Bunnell 1997a, 1998; Chapter 5). We do not know the best way to monitor some elements of definitions of biological diversity, and must accept interim surrogates, such as species or populations for genetic variation. Where our knowledge is sufficient to monitor processes governing outcomes or specific threats to desired outcomes, these should be monitored as well. Desired outcomes must be monitored along with the activities performed to attain those outcomes. Individual forest practices are complex, and effects must be integrated into larger variables that are believed to link directly with the outcomes. Useful candidates include rate of forest removal by all causes; frequency distributions of patch size and age-class distributions for major forest types (the utility of this metric is still poorly documented [Bunnell et al. 1997]); amounts of edge; and approximate abundance of significant habitat elements such as snags, downed wood, and shrubs. Some of these variables can be obtained from map or air photo interpretation. These and other variables, specified below, should be related through adaptive management to the four broad goals or outcomes encompassed by a commitment to maintaining biological diversity (e.g., question 1). In many instances, links between variables summarizing forest activities and outcomes are poorly quantified, and the monitoring should be accompanied by more intensive research.

In terms of the four public goals listed under question 1 ("What Is the Goal?"), monitoring should take the following into consideration.

(a) *Reduce rates of extinction.* Although the fundamental units of diversity are genetic, species are the most recognizable integrated units of genetic diversity (Chapter 3), and monitoring genetic diversity encounters an apparent paradox. Briefly, the paradox is this: the same processes (such as isolation) that help to create species and increase genetic diversity can also reduce existing genetic richness and lead to local extinction. The issue is one of time frame; speciation requires long periods while local extinction can happen quickly.[3]

Currently we have little guidance on the potential outcomes of isolation for different species (e.g., Slatkin 1987). The topic is fruitful for research but provides little guidance for monitoring. Moreover, the short-term threat appears to be isolation of small populations leading to local extirpation (e.g., Chapter 4), not reduced rates of speciation. Species themselves inhabit or are found in a range of habitats, communities, or ecosystems, but these entities are much less fixed than are species (Chapter 5). Direct linkages to forest practices can be encouraged by invoking three objectives:

- maintain habitats of critical importance to particular species
- maintain old-growth elements and some forest-interior habitat
- use forest management techniques (for example, irregular thinning) that maintain stand structures, species composition, and landscape patterns that do not depart markedly from those generated by natural disturbances.

Evaluating the success of reducing rates of extinction can be done by using these objectives to link species presence with the candidate variables summarizing forest activities, such as rate of cut or density of large snags. Evaluating to monitor success should occur hierarchically (e.g., Scientific Panel 1995). Current knowledge suggests that species checklists relating presence or absence to candidate variables is a credible starting point. A growing body of literature on the approach (e.g., Kremen 1992; Green and Young 1995) can be adapted to local forest conditions.

Problems with checklists were reviewed by the Scientific Panel (1995). The panel concluded:

> Despite their limitations, checklists provide an overview of species diversity. When applied at the site level in conjunction with habitat measurements they indicate relations among monitored habitat elements (e.g., snags, downed wood) and individual species. When compared across a range of sites they help to address how much of each element should be retained. At the watershed and site levels they help to evaluate the role of reserves in maintaining species and permit evaluation of the distribution of forest type and age class, including forest patch size and potential edge effects. At the watershed and larger levels the interpretation of repeated and well-distributed site-level checklists can reveal growing "holes" or empty areas in a species range, thus potential problems.

---

3 The apparent paradox emphasizes the importance of distinguishing interim operational targets for sustaining biological diversity from the scientific concepts embraced by the term.

The latter point about expanding empty areas within a species range recognizes the importance of connections among subpopulations of some species (Chapter 4). Unfortunately, it also re-encounters the paradox that processes encouraging the development of new species may encourage local extirpation of others. The challenge is to avoid increasing the rate at which naturally small and localized populations become extinct while not decreasing the rate at which similar, but new, populations are established.

(b) *Sustain productive ecosystems.* Beyond maintaining genetic and species diversity, major determinants of productive ecosystems include slope stability, soil health, stream flow, and water quality. Key indicators for monitoring slope stability are the number of failures and volume of soil displaced per unit time. For soil health, key indicators are forest floor cover (or the obverse, mineral soil exposure), internodal growth, or foliar nutrients. These indices are themselves integrative, serving to some extent as bioassays. The advantage of these indicators is their ease of measurement and known direct relationships to other variables of interest such as tree and forest productivity or likelihood of erosion. Indicators of forest soil health are best applied and evaluated at the site or treatment unit level. Indicators of slope stability or hydrological outcomes must encompass larger areas. Several works provide specific suggestions for monitoring elements of watershed integrity such as hillslopes, stream channels, stream flow, and water quality (e.g., Gregory et al. 1991; Whiting and Bradley 1993; Scientific Panel 1995).

(c) *Retain future options.* Retaining future options is best addressed at two very different scales. The first is the maintenance of productive ecosystems using indices such as mineral soil exposure that are best applied at the finer scales of the physical hierarchy: site (treatment unit) or watershed levels. A second set of outcomes (sustaining all species or economic opportunities) can be evaluated only over larger areas or higher levels of hierarchy. Using broader scales acknowledges that communities inevitably experience species turnover (Chapter 5) and respond to changes in structure by succession and management, and that estimating sustainable fibre production must occur over large enough areas to encompass forest tracts of all ages up to the appropriate rotation length.

Currently the only way of estimating the likelihood of retaining future options over this broader scale is by judicious use of computer-based decision-support systems that attempt to incorporate relevant biological relationships. These systems explicitly relate the candidate forest variables to species responses over space and time (e.g., Hansen et al. 1993; Daust and Bunnell 1994; Daust et al. 1994; Wells et al. 1997). They thus attempt to project the likelihood of sustaining species while simultaneously projecting the volume of the timber harvest (and thus jobs and economic opportunities). Given the relative novelty of using larger spatial scales for research activities, these systems are currently better developed as research frameworks than as planning tools. They are thus most useful when embedded into an adaptive management and policy approach. Their usefulness to monitoring is that they provide an initial framework in which variables describing forest activities are related to desired outcomes. This framework typically contains conjectures and assumptions that can be evaluated by monitoring.

(d) *Retain economic opportunities.* To a great extent this goal is indirectly monitored by indices associated with the three preceding goals. Direct means are familiar and include employment and long-term projected harvest or timber supply.

(16) How Do We Monitor?

We do this by comparing conditions on altered and unaltered sites, before alteration and over long periods afterwards, using simple, credible measurements.

Detecting the environmental impacts of human activities on natural communities is a central problem in applied ecology (Eberhardt 1976; Schroeter et al. 1993). It is a difficult problem because the effects of human activities must be separated from the considerable variability displayed by any ecosystem in the absence of human intervention. Baseline data, collected before the planned intervention, are therefore critical and their collection may rely on the existence of protected areas. Deciding when and how often to carry out post-management inspections requires judgment. Unwanted effects should be detected early enough that changes can be made before damage is done. Some effects of forest practices, however, take time to appear. For example, the roots of harvested trees take time to decay but their decay may affect subsurface drainage and associated slope stability. This time lag means that monitoring must be continued for years after logging operations have ended, to ensure that all effects of logging have become apparent. It also suggests the value of retrospective studies.

Monitoring to detect changes caused by forestry activities is best done by comparing conditions at altered and unaltered sites. Such comparisons can be made by *comparing* conditions: (1) upstream and downstream of a disturbed site; (2) before and after disturbance at the same site; and (3) in the valley containing the disturbed site and in a nearby similar, undisturbed valley (Scientific Panel 1995). None of these possibilities is best for all purposes, and informed choices must be made for each specific purpose. The third comparison is subject to charges of pseudoreplication (but see Bunnell and Huggard 1998), and chronosequences (scattered sites of different ages since logging) have often proven unhelpful (review in Bunnell et al. 1997). Consensus is emerging that to detect impacts of human activities, samples should be taken repeatedly and contemporaneously at the potential impact site and one or more control sites during periods before and after the impact has begun (Eberhardt 1976; Skalski and Mackenzie 1982; Carpenter et al. 1989; Stewart-Oaten et al. 1992; Schroeter et al. 1993). The objective is to exclude or identify, as far as possible, effects unrelated to human activities.

For a given monitoring effort (time, cost), it may be appropriate to make detailed and precise measurements – perhaps with expensive equipment requiring expert operators – on only a few occasions at widely spaced observation sites. Alternatively, numerous observations may be made using less precise methods. Less precise methods allow data to be collected cheaply, at many sites and at short intervals, by observers who have had only minimal training. The ideal is likely some blend of both approaches, allowing simple but extensive observations to be periodically calibrated against complex, intensive measurements. For topographically diverse settings, environmental conditions vary so markedly over such short distances that numerous observation points are necessary. It is therefore worthwhile to devote considerable effort to devising the simplest methods that will give informative results. A fundamental issue is that extensive monitoring must focus on integrative surrogate measures while more intensive study determines the reliability of the surrogates (questions 1 and 15). For example, stream turbidity can be assessed cheaply and extensively, but it is unclear whether it is sufficient to assess slope stability and terrestrial riparian diversity.

(17) How Do We Make Adaptive Management Work?
We do this by clearly specifying goals, monitoring actions intended to attain those goals, and changing course when we have guessed wrongly.

Active adaptation is especially important when information is incomplete and social goals are changing – a situation particularly true of managing to sustain biological diversity. Adaptive management sounds wonderfully attractive, if only because it is the way we learned to walk, talk, ride a bike, and do everything else. We try, note our mistakes, and try again. Despite this innate appeal, implementation of adaptive management has a dismal track record (e.g., Lee 1993; Ludwig et al. 1993). We can and do err at all three steps. Without clear goals, there is nothing against which to evaluate success or progress. Monitoring the results of actions intended to achieve goals requires, at minimum, a commitment of funds and effort to assess the effectiveness of actions. Nowhere have funds commensurate with the values expected from forests been committed to the monitoring of forest practices (Bunnell et al. 1997). Adaptive management may also require careful attention to the manner in which management actions are implemented; these actions are treated as an experiment, and interpretation of their results may require thoughtful experimental design (Walters 1986). Despite the thought and funds required for these steps, we fail most often by not changing course when evidence suggests that we should. It was for this reason that the Clayoquot Scientific Panel recommended that policy as well as management should be adaptive (Scientific Panel 1995).

Insufficient funds and insufficient reflection can each undermine monitoring activities. Research to devise the simplest (cheapest) methods that will give informative results is important (question 16). It is equally important to recall that we do not evaluate the large cost of failing to learn only because we do not know how. Despite the complexity embraced by biological diversity, clear goals can be specified that relate directly to biological diversity (question 1) and the pursuit of natural disturbance regimes as a potential target (question 12). Approaches to learning more from management actions are being developed (e.g., Walters et al. 1988; Walters and Holling 1990) and should be implemented. Failure to change course in the face of compelling evidence is a more difficult issue. It has two large components: psychological and institutional.

Despite the risk inherent in any management action in a complex environment, managers rarely discuss risks openly. An environment receptive to change cannot be created without such a discussion. The main psychological barriers appear to be the inherent fear of publicly owning a mistake and the gratification of being granted control (Bunnell 1976). Confessing one's incomplete knowledge is an acknowledgment of the complexity of forestry, avoids ethical conundrums of perceived control where such are unwarranted (Bunnell 1991), and should contribute to a shared sense of risk taking and thus greater freedom of practice.

Institutional barriers are at least as formidable. They formalize psychological barriers and are ubiquitous enough to be captured in an aphorism: "Big machines move slowly." Many of these barriers have been discussed by Lee (1993) and Ludwig et al. (1993). An obvious and critical first step is restoring the humility and honesty required for practitioners and their critics alike to mutually observe that "we aren't quite sure what we're up to." This caution has been addressed directly by some authors (e.g., Maini in Chapter 2) through the precautionary principle, usually directed at practitioners. It applies equally to government agencies that invoke specific guidelines (implemented as rules) for practice. Landscapes once implemented by forest practice require at least one rotation to modify, so major changes to practice should not be applied universally unless we are very confident of the outcome.

I draw four important points from the chapters in this volume:

(1) We can identify the major public and scientific concerns embraced by ill-defined terms such as *biodiversity* or *natural disturbance regime*. We need not await tedious scientific disclosure of the entire scientific content of such terms.

(2) We can relate our concerns to applicable knowledge and use short-term steering to attain long-term goals. The corollary is that we must recognize such steering as short-term, monitor, and be prepared to change course.

(3) Elements of forest structure and natural disturbance can be useful surrogates for the full content of biological diversity in guiding plans and actions. Our challenge, however, is not to define or to mimic natural disturbances but to understand how disturbance affects elements of biological diversity.

(4) Finally, whatever we do, it must include a variety of approaches to create a wide range of forest patterns. Experience will guide the approaches but only monitoring will refine them.

Literature Cited

Angelstam, P., and G. Mikusinski. 1994. Woodpecker assemblages in natural and managed boreal and hemiboreal forest – a review. Annales Zoologica Fennici 31:157-72.

Art, H.W. (ed.). 1993. The dictionary of ecology and environmental science. Henry Holt, New York, NY.

Avise, J.C. 1995. Mitochondrial DNA polymorphism and a connection between genetics and demography of relevance to conservation. Conservation Biology 9:686-90.

BC Forest Service. 1995. Biodiversity guidebook: Forest Practices Code of British Columbia. BC Forest Service, Victoria, BC.

Binkley, C.S. 1992. Forestry after the end of nature. Forestry Chronicle 69:33-37.

–. 1997. Preserving nature through intensive plantation forestry: The case for forestland allocation with illustrations from British Columbia. Forestry Chronicle 73:553-59.

Botkin, D.B. 1990. Discordant harmonies. A new ecology for the twenty-first century. Oxford University Press, New York, NY.

Bunnell, F.L. 1976. The myth of the omniscient forester. Forestry Chronicle 52(3):150-52.

–. 1989. Alchemy and uncertainty: What good are models? USDA Forest Service, Pacific Northwest Research Station, General Technical Report PNW-GTR-232, Portland, OR.

–. 1990. Biodiversity: What, where, why, and how. Pp. 29-45 *in* A. Chambers (ed.). Wildlife-forestry symposium, Prince George, BC. FRDA Report 160, BC Ministry of Forests/Forestry Canada, Victoria, BC.

–. 1991. Painted in a corner. Pp. 1-6 *in* Ethical challenges for foresters. Proceedings of a symposium at the University of British Columbia, Faculty of Forestry, Vancouver, BC.

–. 1992. De mo' beta blues: Coping with the landscape. Pp. 45-58 *in* Proceedings of a seminar on Integrated Resource Management. Information Report M-X-183E/F, Forestry Canada, Fredericton, NB.

–. 1994. Toto, this isn't Kansas: Changes in integrated forest management. Pp. 1-14 *in* Selected papers from the integrated forest management workshop. Canadian Model Forest Program, Ottawa, ON.

–. 1995. Forest-dwelling vertebrate faunas and natural fire regimes in British Columbia: Patterns and implications for conservation. Conservation Biology 9:636-44.

–. 1996. Forest issues in BC: The implications of zoning. Truck Logger 19(3):49-55.

–. 1997a. Operational criteria for sustainable forestry: Focussing on the essence. Forestry Chronicle 73:679-84.

–. 1997b. Buzzwords, buzzsaws, and buzzards: Refining directions for the 21st Century. 1997 Starker Lectures. Oregon State University. In press.

–. 1998. Overcoming paralysis by complexity when establishing operational goals for biodiversity. Journal of Sustainable Forestry 7:145-64.

Bunnell, F.L., and A. Allaye-Chan. 1984. Potential of ungulate winter-range reserves as habitat for cavity-nesting birds. Pp. 357-65 *in* W.R. Meehan, T.R. Merrell, Jr., and T.A.

Hanley (eds.). Proceedings of a symposium on fish and wildlife relationships in old-growth forests, 12-15 April 1982, Juneau, Alaska. American Institute of Fisheries, Alaska District.

Bunnell, F.L., and L.A. Dupuis. 1994. Canadian-based literature: Implications to conservation and management. EcoScience 1:87-92.

–. 1995a. Riparian habitats in British Columbia: Their nature and role. Pp. 7-21 *in* K. Morgan and M.A. Lashmar (eds.). Riparian habitat management and research. Special Publication of the Fraser River Action Plan, Canadian Wildlife Service, Ladner, BC.

–. 1995b. *Conservation Biology*'s literature revisited: Wine or vinaigrette? Wildlife Society Bulletin 23:56-62.

Bunnell, F.L., and D.J. Huggard. 1998. Biodiversity across spatial and temporal scales: Problems and opportunities. Forest Ecology and Management. In press.

Bunnell, F.L., and L.L. Kremsater. 1994. Tactics for maintaining biodiversity in forested ecosystems. Pp. 62-72 *in* I. Thompson (ed.). Proceedings of the XXI International Union of Game Biologists Congress, Vol. 1.

Bunnell, F.L., and R.G. Williams. 1980. Subspecies and diversity – the spice of life or prophet of doom. Pp. 246-57 *in* R. Stace-Smith, L. Johns, and P. Joslin (eds.). Threatened and endangered species and environments in British Columbia and the Yukon. BC Ministry of Environment, Victoria, BC

Bunnell, F.L., D.K. Daust, W. Klenner, L.L. Kremsater, and R. McCann. 1991. Managing for biodiversity in forested ecosystems. Report to the Forest Sector of the Old-Growth Strategy. Centre for Applied Conservation Biology, University of British Columbia, Vancouver, BC.

Bunnell, F.L., C. Galindo-Leal, and J. Nelson. 1993. Ecological restoration in forested landscapes: Problems and opportunities (British Columbia). Restoration & Management Notes 1:56-57.

Bunnell, F.L., L.L. Kremsater, and R. Wells. 1997. Likely consequences of forest management on terrestrial, forest-dwelling vertebrates in Oregon. Centre for Applied Conservation Biology, University of British Columbia, Vancouver, BC.

Burgman, M.A., S. Ferson, and H.R. Akcakaya. 1993. Risk assessment in conservation biology. Chapman and Hall, London, UK.

Carpenter, S.T., T.M. Frost, D. Heisey, and T.K. Kratz. 1989. Randomization intervention analysis and the interpretation of whole-ecosystem experiments. Ecology 70:1142-52.

Clark, J.S. 1990. Fire and climate change the last 650 years in northwestern Minnesota. Ecological Monographs 60:135-59.

Cline, S.P., A.B. Berg, and H.M. Wight. 1980. Snag characteristics and dynamics in Douglas-fir forests, western Oregon. Journal of Wildlife Management 44:773-86.

Cooperrider, A. 1991. Conservation of biodiversity on western rangelands. Pp. 40-53 *in* W.E. Hudson (ed.). Landscape linkages and biodiversity. Island Press, Washington, DC.

Cumming, S.G., P.J. Burton, and B. Klinkenberg. 1996. Boreal mixedwood forests may have no "representative" areas: Some implications for reserve design. Ecography 19:162-80.

Daust, D.K., and F.L. Bunnell. 1994. Geographic information systems and forest wildlife: Recent developments and future prospects. Pp. 348-56 *in* I. Thompson (ed.). Proceedings of the XXI International Union of Game Biologists Congress, Vol. 1.

Daust, D.K., C. Galindo-Leal, and F.L. Bunnell. 1994. Predicting impacts of forest management on biological diversity. FRDA Research Memo No. 221, BC Ministry of Forests, Victoria, BC.

Delong, D.C. 1996. Defining biodiversity. Wildlife Society Bulletin 24:738-49.

Eberhardt, L.L. 1969. Some aspects of species diversity models. Ecology 50:503-5.

–. 1976. Quantitative ecology and impact assessment. Journal of Environmental Management 4:27-70.

Ehrenfeld, D. 1992. Ecosystem health and ecological theory. Pp. 135-43 *in* R. Constanza, B.G. Norton, and B.D. Haskell (eds.). Ecosystem health: New goals for environmental management. Island Press, Washington, DC.

Ehrlich, P., and A. Ehrlich. 1981. Extinction: The causes and consequences of the disappearance of species. Random House, New York, NY.

Erwin, T.L. 1991. An evolutionary basis for conservation strategies. Science 252:750-52.

FEMAT (Forest Ecosystem Management Assessment Team). 1993. Forest ecosystem management: An ecological, economic, and social assessment. Report of the Forest Ecosystem Management Assessment Team. Portland, OR.

Frankel, O.H., and M.E. Soulé. 1981. Conservation and evolution. Cambridge University Press, Cambridge, UK.

Frankham, R. 1995. Effective population size/adult population size ratios in wildlife: A review. Genetic Research, Cambridge 66:95-107.

Franklin, J.F. 1988. Structural and functional diversity in temperate forests. Pp. 166-75 *in* E.O. Wilson (ed.). Biodiversity. National Academy Press, Washington, DC.

–. 1993. Preserving biodiversity: Species, ecosystems, or landscapes? Ecological Applications 3:202-5.

Franklin, J.F., D.A. Perry, T.D. Schowalter, M.E. Harmon, A. McKee, and T.A. Spies. 1989. Importance of ecological diversity in maintaining long-term site productivity. Pp. 83-96 *in* D.A. Perry, R. Meurisse, B. Thomas, R. Miller, J. Boyle, and R.F. Powers (eds.). Maintaining the long-term productivity of Pacific Northwest forest ecosystems. Timber Press, Portland, OR.

Frehlich, L.E., and P.B. Reich. 1995. Neighbourhood effects, disturbance, and succession in forests of the western Great Lakes region. EcoScience 2:148-58.

Galindo-Leal, C., and F.L. Bunnell. 1995. Ecosystem management: Implications and opportunities of a new paradigm. Forestry Chronicle 71:601-6.

Gibeau, M.L. 1995. Grizzly bear habitat effectiveness model for Banff, Yoho and Kootenay National Parks, Canada. International Conference of Bear Research and Management 10. In press.

Green, R.H., and R.C. Young. 1995. Sampling to detect rare species. Ecological Applications 3:351-56.

Gregory, S.V., F.J. Swanson, W.A. McKee, and K.W. Cummins. 1991. An ecosystem perspective of riparian zones. BioScience 41:540-50.

Grumbine, R.E. 1994. What is ecosystem management? Conservation Biology 8:27-38.

–. 1997. Reflections on "What is ecosystem management?" Conservation Biology 11:41-47.

Hagar, J.C., W.C. McComb, and C.C. Chambers. 1995. Effects of forest practices on wildlife. Chapter 9 *in* R.L. Beschta, J.R. Boyle, C.C. Chambers et al. (eds.). Cumulative effects of forest practices in Oregon: Literature and synthesis. Oregon State University, Corvallis, OR.

Haila, Y., and R. Levins. 1995. Humanity and nature. Ecology, science and society. Pluto Press, London, UK.

Hansen, A.J., S.L. Garman, B. Marks, and D.L. Urban. 1993. An approach for managing diversity across multiple-use landscapes. Ecological Applications 3:481-96.

Hebda, R.J. 1997. Impact of climate change on biogeoclimatic zones of British Columbia and Yukon. Pp. 13.1-13.15 *in* E. Taylor and B. Taylor (eds.). Responding to global climate change in British Columbia and Yukon. Volume 1 of the Canada Country Study: Climate impacts and adaptation. Environment Canada and BC Ministry of Environment, Lands and Parks, Vancouver, BC.

Hengeveld, R. 1992. Right and wrong in ecological explanation. Journal of Biogeography 19:345-47.

Hobbs, R.J. 1992. The role of corridors in conservation: Solution or bandwagon. Trends in Ecology and Evolution 7:389-92.

Hurlbert, S.H. 1971. The nonconcept of species diversity: A critique and alternative parameters. Ecology 52:577-86.

Johnson, K.H., K.A. Vogt, H.J. Clark, et al. 1996. Biodiversity and the productivity and stability of ecosystems. Trends in Ecology and Evolution 11:372-77.

Knight, R.L., and K.J. Gutzwiller (eds.). 1995. Wildlife and recreationists, coexistence through management and research. Island Press, Washington, DC.

Kremen, C. 1992. Assessing the indicator properties of species assemblages for natural areas monitoring. Ecological Applications 2:203-17.

Lapin, M., and B.V. Barnes. 1995. Using the landscape ecosystem approach to assess species and ecosystem diversity. Conservation Biology 9:1148-58.

Lee, K.N. 1993. Compass and gyroscope: Integrating science and politics for the environment. Island Press, Washington, DC.

Lesica, P., and F.W. Allendorf. 1995. When are peripheral populations valuable for conservation? Conservation Biology 9:753-60.

Levin, S.A. 1992. The problem of pattern and scale in ecology. Ecology 73:1943-76.

Levins, R. 1970. Extinction. Pp. 77-107 *in* M. Gerstenhaber (ed.). Lectures on mathematics in the life sciences. Vol. 2. American Mathematical Society, Providence, RI.

Ludwig, D., R. Hilborn, and C. Walters. 1993. Uncertainty, resource exploitation, and conservation: Lessons from history. Science 160:17-36.

MacArthur, R.H., and E.O. Wilson. 1967. The theory of island biogeography. Princeton University Press, Princeton, NJ.

Magurran, A.E. 1988. Ecological diversity and its measurement. Princeton University Press, Princeton, NJ.

Mayr, E. 1963. Animal species and evolution. Harvard University Press, Cambridge, MA.

McAllister, D. 1991. What is biodiversity? Canadian Biodiversity 1:4-6.

McNeely, J.A., K.R. Miller, W.V. Reid, R.A. Mittermeir, and T.B. Werner. 1990. Conserving the world's biological diversity. International Union for Conservation of Nature and Natural Resources, World Resources Institute, Conservation International, World Wildlife Fund (US), and the World Bank. Gland, Switzerland and Washington, DC.

McPherson, M.F. 1985. Critical assessment of the value of and concern for the maintenance of biological diversity. Pp. 154-245 *in* Office of Technology Assessment. 1986. Technologies to maintain biological diversity. Volume 2, Contract Papers. Part E. Valuation of biological diversity. US Congress, Washington, DC.

Naeem, S., J. Thompson, S.P. Lawler, J.H. Lawton, and R.M. Woodfin. 1994. Declining biodiversity can alter the performance of ecosystems. Nature 368:734-37.

Namkoong, G. 1991. Biodiversity issues in genetics, forestry, and ethics. Forestry Chronicle 68:438-43.

–. 1993. Integrating science and management at the University of British Columbia. Journal of Forestry 91:24-27.

Newmark, W.D. 1995. Extinction of mammal populations in western North American parks. Conservation Biology 9:512-26.

Norton, B.G. 1986. On the inherent danger of undervaluing species. Pp. 110-37 *in* B.G. Norton (ed.). The preservation of species: The value of biological diversity. Princeton University Press, Princeton, NJ.

NSF (National Science Foundation). 1989. Loss of biodiversity: A global crisis requiring international solutions. A report to the National Science Board, Committee on International Science's Task Force on Global Biodiversity. National Science Foundation, Washington, DC.

OTA (Office of Technology Assessment). 1987. Technologies to maintain biological diversity. Summary. Report OTA-F-330. Government Printing Office, Washington, DC.

Ohmann, J.L., W.C. McComb, and A.A. Zumrawi. 1994. Snag abundance for primary cavity nesting birds on nonfederal lands in Oregon and Washington. Wildlife Society Bulletin 22:607-20.

Peet, R.K. 1974. The measurement of species diversity. Annual Review of Ecology and Systematics 5:285-307.

Peters, R.H. 1991. A critique for ecology. Cambridge University Press, Cambridge, UK.

Pielou, E.C. 1975. Ecological diversity. John Wiley and Sons, New York, NY.

–. 1995. Biodiversity versus old-style diversity: Measuring biodiversity for conservation. Chapter 3 *in* T.J.B. Boyle and B. Boontawee (eds.). Measuring and monitoring biodiversity in tropical and temperate forests. Proceedings of an IUFRO symposium held at Chiang Mai, Thailand, 27 August to 2 September 1994. Centre for International Forestry Research (CIFOR), Bogor, Indonesia.

Popper, K.R. 1959. The logic of scientific discovery. Hutchinson, London, UK.

Prendergast, J.R., R.M. Quinn, J.H. Lawton, et al. 1993. Rare species, the coincidence of diversity hotspots and conservation strategies. Nature 365:335-37.

Reid, W.W., and K.R. Miller. 1989. Keeping options alive: The scientific basis for conserving biodiversity. World Resources Institute, Center for Policy Research, Washington, DC.

Ryti, R.T. 1992. Effect of the focal taxon on the selection of nature reserves. Ecological Applications 2:404-10.

SAF (Society of American Foresters). 1991. Task force report on biological diversity in forested ecosystems. Bethesda, MD.

Sagoff, M. 1993. Biodiversity and the culture of ecology. Bulletin of the Ecological Society of America 74:374-81.

Samson, F.B., and F.L. Knopf. 1993. Managing for biodiversity. Wildlife Society Bulletin 21:509-14.

Schaeffer, D.J., E.E. Herricks, and H.W. Kerster. 1988. Ecosystem health I: Measuring ecosystem health. Environmental Management 12:444-55.

Scherer, D. (ed.). 1990. Upstream/downstream: Issues in environmental ethics. Temple University Press, Philadelphia, PA.

Schroeter, S.C., J.D. Dixon, J. Kastender, and R.O. Smith. 1993. Detecting the ecological effects of environmental impacts: A case study of kelp forest invertebrates. Ecological Applications 3:331-50.

Scientific Panel for Sustainable Forest Practices in Clayoquot Sound. 1995. Report 5, Sustainable ecosystem management in Clayoquot Sound: Planning and practices. BC Ministry of Environment, Lands and Parks, Victoria, BC.

Scudder, G.G.E. 1989. The adaptive significance of marginal populations: A general perspective. *In* C.D. Levings, L.B. Holtby, and M.A. Henderson (eds.). Proceedings of the national workshop on effects of habitat alteration on salmonid stocks. Canadian Special Publication of Fisheries and Aquatic Sciences 105:180-85.

Selander, R.K. 1976. Genetic variation in natural populations. Pp. 21-45 *in* F.J. Ayala (ed.). Molecular evolution. Sinauer Associates, Sunderland, MA.

Shrader-Frechette, K.S., and E.D. McCoy. 1993. Method in ecology: Strategies for conservation. Cambridge University Press, Cambridge, UK.

–. 1995. Natural landscapes, natural communities, and natural ecosystems. Forest and Conservation History 39:138-42.

Simberloff, D., J.A. Farr, J. Cox, and D.W. Mehlman. 1992. Movement corridors: Conservation bargains or poor investments? Conservation Biology 6:493-504.

Simpson, G.G. 1944. Tempo and mode in evolution. Columbia University Press, New York, NY.

Skalski, J.R., and D.H. Mackenzie. 1982. A design for aquatic monitoring programs. Journal of Environmental Management 14:237-52.

Slatkin, M. 1987. Gene flow and the geographic structure of natural populations. Science 236:787-92.

Slocombe, D.S. 1993. Implementing ecosystem-based management. BioScience 43:612-22.

Soulé, M.E. 1987. Viable populations for conservation. Cambridge University Press, New York, NY.

Spellerberg, I.F. 1992. Evaluation and assessment for conservation: Ecological guidelines for determining priorities for nature. Chapman and Hall, London, UK.

Spies, T.A., J.F. Franklin, and T.B. Thomas. 1988. Coarse woody debris in Douglas-fir forests of western Oregon and Washington. Ecology 69:1689-702.

Stewart-Oaten, A., J.R. Bence, and C.W. Osenberg. 1992. Assessing effects of unreplicated perturbations: No simple solutions. Ecology 73:1396-404.

Swanson, F.J., F.A. Jones, D.O. Wallin, and J.H. Cissel. 1993. Natural variability – implications for ecosystem management. Pp. 89-103 *in* M.E. Jensen and P.S. Bourgeron (eds.). Eastside forest ecosystem health assessment, Vol. 2, Ecosystem management: Principles and applications. USDA Forest Service, Portland, OR.

Takacs, D. 1996. The idea of biodiversity: Philosophies of paradise. Johns Hopkins University Press, Baltimore, MD.

Tilman, D., and A. Downing. 1994. Biodiversity and stability in grasslands. Nature 367:363-65.

Tracy, C.R., and P.F. Brussard. 1994. Preserving biodiversity: Species in landscapes. Ecological Applications 4:205-7.

United Nations. 1992. Convention on biological diversity.

Walker, B.H. 1992. Biodiversity and ecological redundancy. Conservation Biology 6:18-23.

–. 1995. Conserving biological diversity through ecosystem resilience. Conservation Biology 9:747-52.

Walters, C.J. 1986. Adaptive management of renewable resources. Macmillan Publishing, New York, NY.

Walters, C.J., and C.S. Holling. 1990. Large-scale management experiments and learning by doing. Ecology 71:2060-8.

Walters, C.J., J.S. Collie, and T. Webb. 1988. Experimental designs for estimating transient responses to management disturbances. Canadian Journal of Fisheries and Aquatic Sciences 45:530-38.

Wells, R., P. Vernier, and G. Sutherland. 1997. Assessing the effects of forest harvesting on landscape patterns and wildlife habitat: Current approaches and new directions for SIMFOR. In press.

Whiting, P.J., and J.B. Bradley. 1993. A process-based classification system for headwater streams. Earth Surface Processes and Landforms 18:603-12.

Whittaker, R.H. 1970. Communities and ecosystems. Macmillan, New York, NY.

Wielgus, R.B., and F.L. Bunnell. 1994. Dynamics of a small, hunted brown bear *Ursus arctos* population in southwestern Alberta. Biological Conservation 67:161-66.

Wilcove, D. 1994. Response. Ecological Applications 4:207-8.

Wilson, E.O. 1995. Naturalist. Warner Books, New York, NY.

Wilson, E.O., and F.M. Peter (eds.). 1988. Biodiversity. National Academy Press, Washington, DC.

Wood, P. 1997. Biodiversity as the source of biological resources: A new look at biodiversity values. Environmental Values 6:251-68.

Yanchuk, A.D., and D.T. Lester. 1996. Setting priorities for conservation of the conifer genetic resources of British Columbia. Forestry Chronicle 72:406-15.

Contributors

Clark S. Binkley was Dean of the Faculty of Forestry at the University of British Columbia from 1990 to 1998. His principal area of research is the application of economics to problems arising in public and private management of forests. He has served as a consultant to a large number of forest products companies, governmental agencies, and private conservation groups, and has served on boards of directors of many committees, agencies, and companies. Dr. Binkley holds a Ph.D. from Yale University. He is presently Senior Vice President, Investment Strategy and Research, Hancock Natural Resource Group, Boston.

Fred L. Bunnell is Forest Renewal Professor in Applied Conservation Biology in the Faculty of Forestry and Director of the Centre for Applied Conservation Biology, Faculty of Forestry, University of British Columbia. He has received several awards including the Bill Young Award, which recognizes excellence in integrated resource management, the Canadian Institute of Forestry Scientific Achievement Award, and the Northwest Science Association's Outstanding Scientist Award.

Ann Chan-McLeod is a Research Associate with the Centre for Applied Conservation Biology at the University of British Columbia. She received her Ph.D. from the University of Alaska Fairbanks. Her research focuses on wildlife physiological ecology, animal dispersal, and conservation of biological diversity in forested ecosystems.

Jagmohan S. Maini is presently Coordinator and Head of the Secretariat for the Intergovernmental Forum on Forests at the United Nations. He has held several Government of Canada posts and has been active in conducting research and developing policy on forestry and the environment both nationally and internationally. Dr. Maini was recently elected Chair of the Commonwealth Forestry Association and is a member of the Board of Trustees, Centre for International Forest Research, Bogor, Indonesia.

Gray Merriam is Professor Emeritus at Carleton University in Ottawa. He was a founding member and past President of the International Association of Landscape Ecology. Throughout his career, he has worked to link theoretical and empirical findings in ecology to policy and management applications.

Gene Namkoong is Professor and former Head, Department of Forest Sciences, University of British Columbia. He serves on the Board of Trustees of the International Plant Genetic Resources Institute and is a member of the United Nations Panel of Experts on Forest Genetics, the Swedish Academy of Agriculture and Forestry, and the Korean Academy of Science and Technology. Among his many honours, he received the Wallenberg Prize from the King of Sweden in 1994, the highest honour in the world awarded to a forest researcher.

Reed F. Noss is co-executive director of the Conservation Biology Institute in Corvallis, Oregon. He is former editor of *Conservation Biology* and president-elect of the Society for Conservation Biology.

J. Stan Rowe is Professor Emeritus, University of Saskatchewan, specializing in Ecology. His interests are landscape ecology and environmental ethics. He is Trustee Emeritus of the Canadian Parks and Wilderness Society.

Daniel Simberloff is Nancy Gore Hunger Professor of Environmental Studies, Department of Ecology and Evolutionary Biology, University of Tennessee. His specializations are conservation biology, invasion biology, biogeography, and community ecology. He is North American editor of *Biodiversity and Conservation* and on the editorial boards of *American Naturalist, Oecologia,* and *Bioscience.*

Index

Adaptive management
  definition, 78
  and public policy, 14
  use of, 135, 146-8
Agenda 21, 6
"Airplane passengers" theory, 122, 125
"Airplane rivets" theory, 122, 125
Alleles
  rare, and need for redundancy, 37
  saving by genetic management, 35
Alpha diversity
  and conservation tactics, 141
  critique of, 11, 13, 131
  definition, 68
  and equality of species question, 124-5
  as index of ecological diversity, 10
  and management tactics, 132
  as measuring tool, 88
  and spatial scale, 105
Animals. *See* Birds; Mammals; names of
  specific animals; Species

"Backcasting," and management programs, 57
Bacteria, nitrogen-fixing, 4
Barren ground caribou, 102
Bay checkerspot butterfly (*Eukphydryas editha bayensis*), 103
BC Ministry of Environment, Lands and Parks, 85
Bensoniella oregana, 100
Beta diversity
  and conservation tactics, 141
  critique of, 11, 13
  definition, 68
  as index of ecological diversity, 10
  and management tactics, 132
  as measuring tool, 88
  and spatial scale, 105

Biodiversity
  and clearcutting, vii
  and community stability, 71
  compositional, 73-4
  definition, 6, 7, 8, 69-70, 118, 131-2
  and dynamic systems, 135-6
  functional, 74
  and genetic diversity, 12, 122
  importance of, 71
  and landscape preservation, 82-94
  and managed forests, 3-7
  mapping projects, 98
  mechanistic models of, 77-8
  origins of term, xiii
  and productivity, in forest systems, 71
  public concern for, vii, 6-7
  and public policy, 7, 20-2, 57-8
  scales for management of, 12, 96-113
  scientific approaches to, ix, 45-7
  strategy development for, 12, 117-48
  See also Alpha diversity; Beta diversity;
    Gamma diversity; Measurement, of
    biodiversity; Monitoring, of
    biodiversity; Values
Biodiversity Guidebook (of Forest
  Practices Code of BC), 139-40
Biogeoclimatic Ecosystem Classification
  (BEC) system, 84-5
Biological Fallacy, 83-4
Biota, and landform formation, 89
Birds, forest-dwelling, 4
Bison, 102
Breeding
  and genetic management, 42
  intensive, 37
British Columbia. *See* BC
Brundtland Commission (World Commission on Environment and Development) report, xiii, 6, 21

9263